ISDN

How to Get a High-Speed Connection to the Internet

Charles Summers
Bryant Dunetz

John Wiley & Sons, Inc.

New York • Chichester • Brisbane • Toronto • Singapore

Publisher: Katherine Schowalter
Editor: Tim Ryan
Assistant Editor: Allison Roarty
Managing Editor: Mark Hayden
Text Design & Composition: Benchmark Productions, Inc.

This text is printed on acid-free paper.

Library of Congress Cataloging-in-Publication Data

Summers, Charles K.
ISDN : how to get a high-speed connection to the Internet / Charles Summers, Bryant Dunetz.
p. cm.
Includes index.
ISBN 0-471-13326-4 (pbk. : alk. paper)
1. Internet (Computer network) 2. Integrated services digital networks. I. Dunetz, Bryant. II. Title.
TK5105.875.I57S86 1996
004.6'7—dc20 95-45824
CIP

Printed in the United States of America

10 9 8 7 6 5 4 3 2 1

To
Cathy, Karl, Cindy
and
my fellow workers
at
TeleSoft International, Inc

Charles K. Summers

To
Gloria, Rick, Jason, Ray
and especially
Kevin
without whom this project would not be possible.

A special thank you to CICAT Networks and to all the service providers and equipment manufacturers who provided information on ISDN and the Internet. I also want to express my appreciation to my co-author who's steady hand of experience in a project such as this was invaluable through the entire production.

Bryant R. Dunetz

CONTENTS

PREFACE

Integrated Services Digital Network (ISDN) is a technology that has been devised to allow people access to much higher data speeds using existing telephone lines. The Internet is a service that forwards, and allows access to, large amounts of data. The technology and service seemed to have been designed for one another. So, why haven't they been? The Internet has been viable for a long time, although its popularity has increased significantly in the past few years. Similarly, ISDN has existed as an implementable concept since about 1988. However, the actual availability of ISDN has been very limited until recently. It is now time for the obvious link to take place.

Many market observers have referred to Internet Access as the so-called "killer application" for ISDN. As the deployment of ISDN telephone service has spread throughout the United States so has the availability of new hardware and new Internet Service Providers offering ISDN connections. Lower prices have also contributed to the increasing popularity. By some estimates, however, there are still less than 50,000 subscribers nationwide connected to the Internet via ISDN. In spite of its slow and cautious beginning its vigorous expansion is imminent. With the roll-out of new residential tariffs by the Local Exchange Carriers and migration to ISDN by the major on-line services, ISDN subscribers to the Internet could reach a million or more by 1998.

To succeed on the scale of modem access to the Internet computer manufacturers will have to include an ISDN feature in their systems and the public will have to learn more about this new technology and its practical features. The purpose of this book is to provide the foundation of knowledge for potential ISDN Internet users as well as an easy to use reference.

CHAPTER

1

CONNECTING TO THE INTERNET VIA ISDN: AN OVERVIEW

NEW DIGITAL INTERFACE TO THE INTERNET

Imagine connecting to the Internet ten times faster than you can using analog modems. You no longer have to imagine because ISDN allows you to do this at speeds of 128,000 Kbps or faster. Because of its speed and flexibility, ISDN is rapidly becoming the Internet access communications medium of choice for individuals and for business. Internet service providers, bulletin board operators, and managed on-line services (for example, Prodigy, CompuServe and America Online), have all realized the importance of ISDN in their marketing strategies. As a result, many are launching major efforts at offering a reasonably priced ISDN interface to their services. ISDN is the fastest dial-up telephone service available today for accessing the Internet and will continue to be the fastest into the foreseeable future.

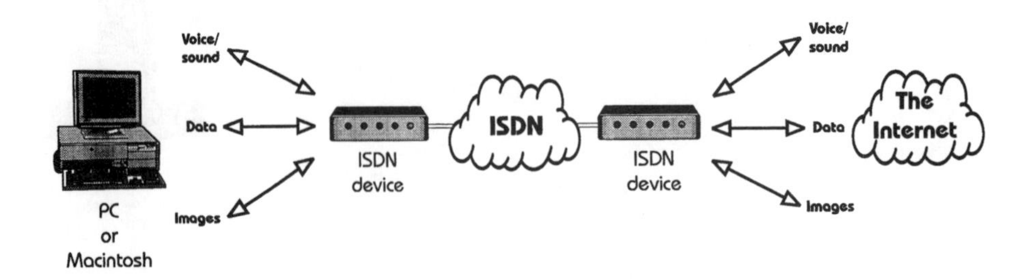

Figure 1.1 Integrated Services Digital Network.

The marriage of ISDN and the Internet is a relatively new occurrence even though ISDN as a technology has been around for more than ten years and the Internet for twenty-five years. By allowing users to send voice, data, and images over existing copper wires (illustrated in Figure 1.1) ISDN allows the panoply of multimedia assets of the Internet to be seen and heard in real time.

Besides its use for high-speed Internet access, ISDN is currently being used in a number of other applications such as remote LAN access, telecommuting, and video conferencing. Rest assured that if you buy an ISDN line to access the Internet, you will be able to use that same line for any of the other ISDN applications, assuming you buy the correct ISDN equipment. One of ISDN's greatest assets is that you can run multiple applications over the same physical connection to the telephone company.

HOW FAST IS ISDN?

Before getting into the technical details of ISDN speed, let's review the measurements of speed and their abbreviations. Bandwidth, speed, and throughput are terms that are frequently used interchangeably when referring to the velocity at which information can move through a communication transmission medium, such as copper wire or fiber optics. Data transmission speeds are commonly measured in *bits per second*. The term *bit* is derived from the smallest unit of digital information, referred to as a binary digit.

ISDN VS. ANALOG TRANSMISSIONS

Assuming a limit of 28.8 Kbps for analog modems, ISDN equivalents will be at least twice as fast and could be as high as five to ten times the speed when high-level compression algorithms are introduced. As new data-intensive applications, such as the World Wide Web and others, are developed for the Internet, the need for increased bandwidth becomes an essential element of performance. For the moment, ISDN is capable of providing that level of performance at 128 Kbps. Figure 1.2 illustrates the relative difference in speed that you can expect from the two types of communications.

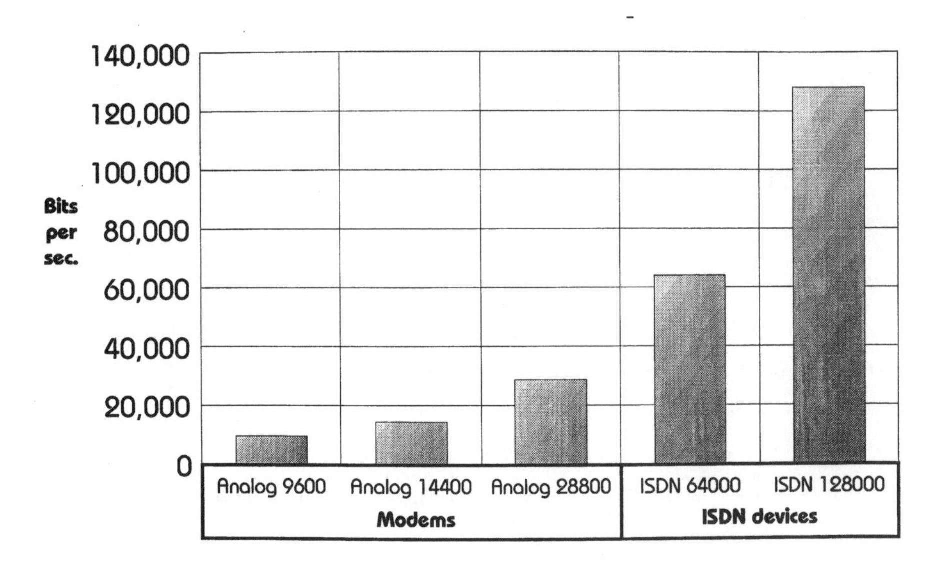

Figure 1.2 Comparison of ISDN and analog modems.

MULTIMEDIA NEEDS ISDN

One of the benefits of marrying ISDN to Internet access is the increase in speed of popular Internet applications, such as World Wide Web, CUSEEME, and FTP. Of the estimated 40 to 50 million Internet users in 1994, approximately 10 million had the capability to use a graphical interface to access the World Wide Web in 1995. When viewed over 14.4 Kbps or 28.8 Kbps modems, the Web can be painfully slow, especially when large graphics or sound files are downloaded. When ISDN is introduced, the increase in speed is obvious as the downloads occur in almost real time.

REALIZING THE SPEED OF ISDN

Because ISDN lines can be linked together in multiples, users will find no theoretical ceiling on the speed of their connection to the Internet. If users can afford two ISDN lines, they will be able to connect to the Internet at 256 Kbps. With the addition of compression technology, ISDN speed can be doubled and tripled as well. ISDN is also fully compatible with other available communications types, including such newer broadband solutions as frame relay and asynchronous transfer mode (ATM). This means that your investment today should be protected into the future as new technologies emerge.

Two to three years ago the modem industry was assessing its future, thinking that without a new market opening, maintaining production targets would be difficult. Sales were level at best, and the Internet and managed on-line services were just becoming known to the public at large. Then in 1993 and 1994, demand exploded as several million customers wanted to get connected—modem sales skyrocketed. At the same time, competition among the leading manufacturers drove modem prices down.

Although 2400 and 9600 baud analog modems were considered the latest and the fastest in the late 1980s and early 1990s, they were quickly replaced with 14.4 Kbps modems. In 1995, the 28.8 Kbps modem was introduced in large numbers; 28.8 Kbps modems, nevertheless, appear to represent industry's theoretical best. Consequently, ISDN becomes the next available replacement for modems and fulfills the present demand for speed and reliability in a communications interface to the Internet.

COMPUTERS ARE MORE POWERFUL

With computer users' desire for "greater capacity," manufacturers seem to find an endless supply of processing power. The current trend today in retail computer sales is heavily weighted toward multimedia, and processor capacity appears to have no limit. Even as they struggle to satisfy market demand for Pentium processors, chip manufacturers are speaking of the next generation of gigaHertz power and speed. This trend should facilitate the utilization and compatibility of ISDN and other forms of high-speed digital communications.

LINKING COMPUTERS AND COMMUNICATIONS

Computers, information, and data networks have become natural bedfellows with the growth in digital communications service and equipment. The modem was the first appliance to allow you to communicate with a remote computer over a standard analog telephone line. PC manufacturers recognized the importance of this relationship and began to provide a slot in their integrated designs for an internal modem. There is no reason to believe that an ISDN interface won't appear in a similar fashion in the mass market. A number of high end and relatively expensive workstations, such as those from Sun and Silicon Graphics, already incorporate an ISDN interface.

WHERE CAN YOU FIND THE TELEPHONE SERVICE?

ISDN telephone service is widely available in the United States, with more than 90 million ISDN-capable circuits available by the end of 1995. Of the

100 million-plus telephone lines installed in the United States, approximately 1 percent of these lines will be ISDN lines by the end of 1995.

ISDN deployments are primarily concentrated in major urban areas however, you can find ISDN in many suburban and rural areas as well. Even though projected deployment numbers are very high, you still may not be able to receive ISDN service. The reasons are: you are not close enough to connect to the ISDN telephone facilities or the service is not being offered by your local telephone company.

ISDN MAY COST MORE THAN YOUR CURRENT SERVICE

ISDN telephone service does not yet compete in price with standard analog service, and it may *never* be as cheap. ISDN telephone service can cost as much as a few hundred dollars to have installed at a business or residence. The monthly charge for ISDN will also be higher than for your analog service. If you need or desire the performance—that is, the speed—ISDN is worth the extra expense.

In addition to the basic rate of approximately $50/month, more or less, for ISDN telephone service, you should also be prepared to pay usage charges (similar to long distance charges) of several cents per minute for data calls. In time, usage rates are expected to come down significantly as the number of subscribers goes up.

INTERNET SERVICE PROVIDERS MIGRATING TO ISDN

Internet service providers nationwide have recognized that ISDN access to the Internet is a critical element of their business strategy in order to meet the demand for faster service. As analog modems are being stacked in almost unmanageable numbers at ISPs around the country, compact and easy-to-manage ISDN hubs are simplifying the logistics of maintaining Internet service. In time, ISDN technology will replace its analog predecessor in many Internet service networks where analog technology can no longer be technically or economically justified.

ISDN access to Internet is available from a number of local and national Internet service providers, starting at prices as low as $30/month for four hours of service per day to $100/month for unlimited access. Most ISPs charge a nominal setup fee, which varies from provider to provider.

According to information made available by the Internet Society, at the beginning of 1995 more than 50 Internet service providers nationwide indicated or advertised that they offer ISDN connections. Because of the relatively new state of the commercial Internet industry, this number could be off by a factor of 2, or more. With the exponential growth in usage of the World Wide Web and other data-intensive applications on the Internet this number of ISDN providers could easily double or triple in 1996. It seems that a day doesn't pass without a new ISDN service provider springing up. Telephone companies, which already have the infrastructure in place, are also taking active positions in the ISDN Internet service market.

Currently the two largest national ISPs are Performance Systems International (PSI) and UUNET. Both companies were early participants in the development of ARPA Net and NSF Net, and they were early users of ISDN communications. The next chapter will furnish additional information on the various types of services available today from local and national Internet service providers.

ISDN HARDWARE EVOLUTION

ISDN hardware has come a long way since the origin of ISDN technology. In the early years, you could only buy an ISDN telephone to use on your ISDN line. ISDN service was being sold as a predominately voice-based service. As companies realized the value of ISDN's data capability, products started popping up that exploited the full voice and data capability of

ISDN. Today, there are many different types of ISDN devices with varying sets of voice and data features. ISDN telephones, video conferencing systems, routers, bridges, computer cards, and terminal adapters (ISDN devices that work like modems) are among the most popular devices. Most of these devices are currently computer peripherals; however, it is conceivable that ISDN interfaces will be built directly into computers when the market is large enough. A number of high-end and relatively expensive workstations such as Sun and Silicon Graphics already incorporate an ISDN interface.

HOW ABOUT THE COST OF ISDN EQUIPMENT?

Typically, ISDN equipment will range from $300, at the low end, for individual Internet users to as much as $2,000 for equipment with advanced features. Even though there is quite a variety of options to choose from, all ISDN equipment has similar basic features, such as the ability to pass data or data and voice. It is also possible to purchase a device with additional communication functions. One option is the number of analog ports pro-

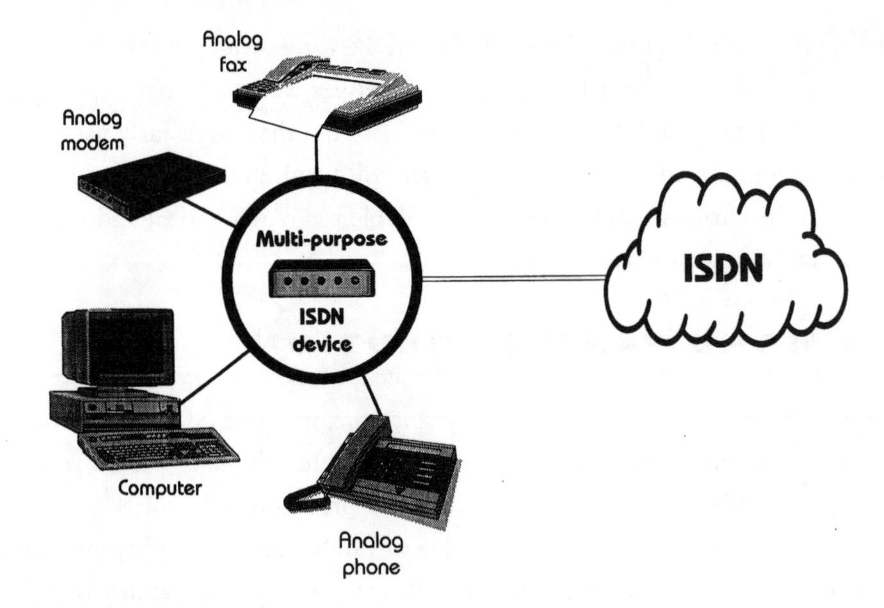

Figure 1.3 Several types of equipment can connect to a single ISDN device.

vided—1 or 2 to support a telephone, fax or modem, or enough ports to support multiple telephones, as shown in Figure 1.3. Whatever the case, additional features generally equate to additional cost. Within the next few years, an ISDN terminal adapter, data-only device, will compete in price with the highest performance 28.8 analog modem.

EQUIPMENT PURCHASES

You will purchase ISDN equipment through your Internet service provider, value-added reseller in your area, or local telephone company. When the business base warrants, you will be able to purchase what you need from the local computer store, just as you now do for analog modems and related computer communications accessories. In fact, ISDN equipment is beginning to show up in retail outlets in selected areas of the country.

A word of caution to new users of ISDN technology is in order. It is very easy to purchase a product which does not have the correct feature for your application. Contact your Internet service provider or equipment specialist for advice on which equipment will give you the best features and performance for your current needs, as well as for your future needs.

LEADING ISDN EQUIPMENT MANUFACTURERS

Definite signs in the industry today substantiate the current growth trend in ISDN. Two formidable leaders in the ISDN business are Ascend Communications and Adtran. Both are growing rapidly—by any measure—to meet present and projected demand for equipment. New products are being introduced by a growing number of vendors competing in what should turn out to be a very lucrative market.

YOU MAY STILL NEED A SPECIALIST TO ASSIST

Because ISDN service and equipment require some knowledge of computers and communications, you may want to consult a specialist on subjects not answered fully. A short consultation with an expert could potentially save you money and lost time in the long run. In addition to contacting a specialist, there are many sources for ISDN information through the Internet or your local telephone company. References to some of this information can be found in the body of each chapter as well as in the Appendixes of this book.

SOME BASICS ABOUT SETTING UP YOUR ISDN INTERNET SERVICE

The process of getting a high-speed connection to the Internet via ISDN can best be described in several general categories:

- Analyzing your requirements
- Obtaining the telephone service
- Securing an ISDN connection from an Internet service provider (ISP)
- Selecting the hardware and software to deliver the best performance

In broad terms, it is satisfactory to set your requirement at "realizing a significant increase in speed over the best available modem." It is also useful to know that you may be shopping for an ISDN connection to the Internet that is twice the speed or four to five times the speed of the analog modem.

Before you embark on a search for an ISP, however, you must determine that ISDN telephone service is available at your residence or office. A call to your telephone company should answer that question.

Finding an ISP can be the one experience in the process which begins to present varied and sometimes complex options that will require explanation and understanding. For now, however, let us assume that you will be looking for an ISP that has a fully developed and widely used ISDN service, and your selection can be made on price alone.

Selecting the hardware and software can be one of the most difficult aspects of getting a connection to the Internet via ISDN. To get the best performance for your Internet access and potentially other ISDN applications, you need to understand how ISDN equipment works and how it interacts with your computer.

THE FUTURE OF ISDN AND THE INTERNET

ISDN is not yet fully appreciated or understood, much like the Internet. Both are relatively new manifestations of modern computer and communications technology, and both are growing in acceptance at an extraordinary pace.

Digital technology is fast replacing analog communications as the preferred method of accessing the information Superhighway.

ISDN high-speed digital communications will open an exciting new dimension of performance for millions of Internet and on-line service users.

Some people have even referred to ISDN as the "on-ramp" to the information superhighway. If you are concerned that ISDN may be replaced by newer or better technologies, you can rest assured because ISDN in its present form will be around for many years to come. This statement is supported by the fact that ISDN is an international standard and billions of dollars have been invested worldwide to deploy the service. In addition, ISDN will interoperate with other technologies, such as frame relay and asynchronous transfer mode (ATM). This means that your investment today will be protected into the future.

SUMMARY

Although not yet at the stage of widespread acceptance, ISDN could conceivably be serving 10 million users across the country, by the year 2000.

Demand for high-speed access to the Internet will certainly be one of the main stimuli for a major migration to ISDN connections. Full national deployment of the telephone service, lower residential tariffs, and solutions for a few nagging technical issues will also contribute to rapid growth.

Presenting this subject to the public in a way that communicates the right message and just enough information, has been the challenge of this book. We hope to address this challenge for the full range of interests characterized as follows:

- Don't bother me with the details just get me a high-speed connection.
- I heard about ISDN. Can someone tell me what it is?
- I just ordered an ISDN line and purchased a terminal adapter. Can you tell me how to connect to the Internet?
- I am considering ISDN for Internet access. Can you describe the procedure to make sure that I get the proper service and equipment?
- I am very familiar with the Internet. Can you tell me how ISDN works using standard Internet protocols and the advantages of utilizing Bonding techniques compared to Multi-Link Point-to-Point Protocol to get 128 Kbps of throughput?

The next 4 chapters are intended to guide you through each step of the process of getting a high-speed connection to the Internet via ISDN. Simply put, they outline where to start when getting an ISDN connection to the Internet and what you should know to realize ISDN's performance. Part 2 of the book addresses the evolution, theory, and standards of ISDN.

PART 1

CHOOSING AND INSTALLING YOUR ISDN SERVICE

CHAPTER

2

ISDN: FOUNDATIONS FOR INTERNET ACCESS

To set yourself up with a high-speed connection to the Internet, you will benefit by understanding what ISDN is and how it works. A basic understanding of the technology will help you avoid conflicts as you navigate through the process of getting connected. Take a few minutes to familiarize yourself with a little bit of the theory behind high-speed performance. A more detailed explanation of the architecture, evolution, protocols, and potential services is provided in the second part of this book.

A BRIEF DESCRIPTION OF ISDN

ISDN is primarily meant to be an international architecture—a method that provides a gradual shifting of the analog communications networks to digital ones. It allows for the use of existing equipment, networks, and services.

It also allows for the possibility of increased services. Thus, ISDN is designed to allow continued use of modems, fax machines, voice terminals, and other services in use within the analog communications world. It does this while providing an increase in the quality and speed of transmission. New services, of course, require newly designed equipment.

WHAT DOES INTEGRATED SERVICES DIGITAL NETWORK MEAN?

The ISDN acronym can be broken into two distinct components: integrated services and digital network. The first component refers to the fact that ISDN can offer many different communication services over the same physical interface to the telephone company. This is different from analog technology where customers need a different physical pair of wires for each different service they might require in their business or home (phone, fax, modem, etc.). With ISDN, the customer can get all of these services over the same two copper wires. ISDN allows customers (via their ISDN device) to specify what service they would like from the telephone network on a call-by-call basis. One call might be a voice call, another call might be a data call. The second component refers to the fact that ISDN is based on digital rather than analog technology. Information is transferred using bits and bytes much as computers operate. This feature makes it very easy for computers to interface to an ISDN network.

TWO LEVELS OF ISDN SERVICE

There are currently two different levels of ISDN service: Basic Rate Interface (BRI) and Primary Rate Interface (PRI). BRI is designed to provide a digital telephone service for individuals. It is therefore the main type of service used to get high-speed access to the Internet. BRI runs over the existing two-wire copper telephone lines that analog telephone service uses; in many cases, you will not have to rewire your home or office to use BRI ISDN. There are three digital channels on a BRI, as depicted in Figure 2.1.

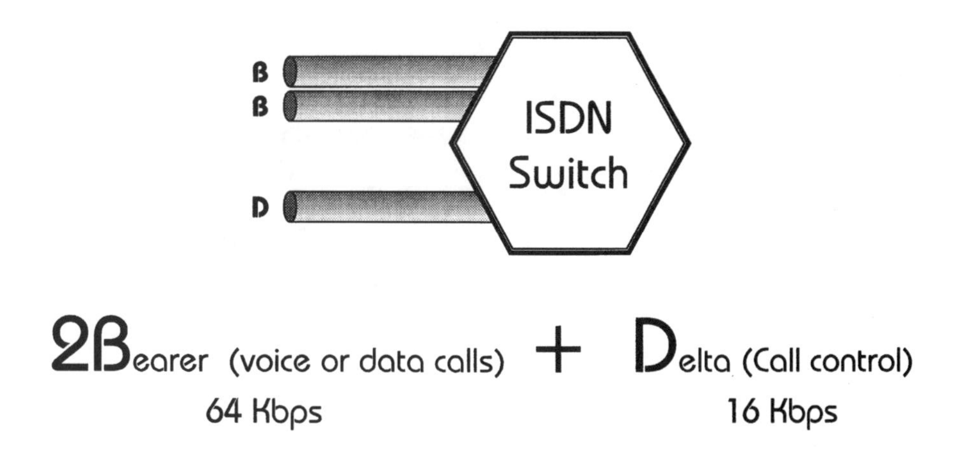

Figure 2.1 Basic Rate Interface (BRI) Single ISDN line for individual users.

Two of the channels are used to carry voice or data calls. These channels are called Bearer (B) channels and have a speed of 64 Kbps each. The third channel is used to communicate control commands to and from the ISDN switch. This channel is called the D-channel (sometimes referred to as a Delta channel to distinguish between it and the B-channel) and has a speed of 16 Kbps. Because there are two B-channels, you can have up to two simultaneous active voice/data calls on a BRI.

PRI was designed to accommodate larger installations where BRI was too small to satisfy the demand. On a PRI, as seen in Figure 2.2, there is a total of 24 digital channels in North America and 31 digital channels in Europe and elsewhere. Normally, 23 (30 outside North America) 64 Kbps channels of a PRI are used for B-channels and one 64 Kbps channel is used for the D-channel. The B-channels are, once again, available for voice and data calls, and the D-channel is used for communication with the ISDN switch. This means that you can work with up to 23 (or 30 outside of North

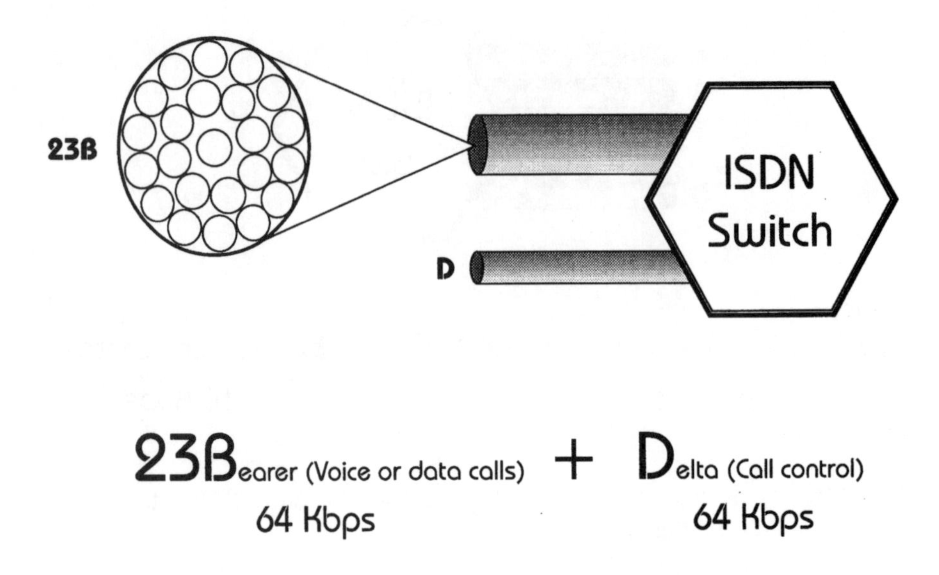

Figure 2.2 ISDN Primary Rate Interface (PRI).

America) simultaneous active voice/data calls (or fewer, but higher-speed, groupings of individual B-channels) on a PRI. A PRI might be used by an Internet provider that has 20 or more customers.

ISDN PROTOCOLS

In the communication world, two devices communicate with each other using what are known as "protocols." Protocols are essentially the same as speaking languages. To get communication devices to talk with each other you need to have common protocols on each end of the connection. Otherwise, the devices will not understand what the other end is saying. If you are familiar with modems, then you may have heard of protocols such as V.32, V.32bis, V.42, V.42bis, and so on. These are the protocols that allow two modems to communicate. When it comes to ISDN protocols, two different categories must be considered: ISDN switch protocols and ISDN B-channel protocols.

ISDN SWITCH PROTOCOLS

The ISDN switch protocol is the language that is spoken over the ISDN D-channel to the ISDN switch. This protocol is designed to control the origination and termination of phone calls, monitor the connection, and provide diagnostics to both the ISDN end user and the ISDN switch. Figure 2.3 demonstrates how this works. There are more than ten different variants of ISDN switch protocols in the world. In the United States, the predominant protocol is called National ISDN-1. The second most popular protocol is called AT&T Custom. EuroISDN (sometimes referred to as EDSS or by the associated test suites, NET3) is considered the standard

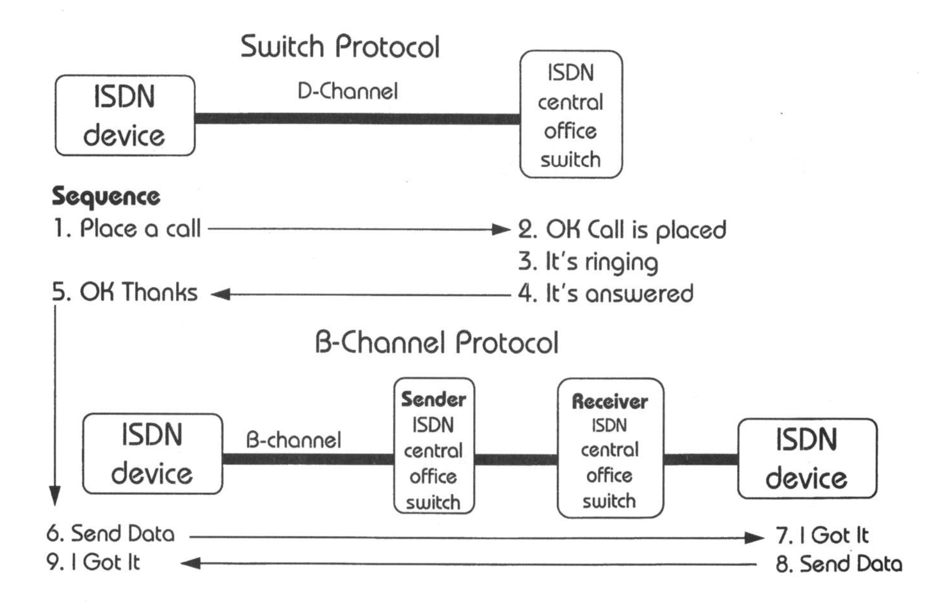

Figure 2.3 ISDN switch and B-channel protocols (how they work).

ISDN protocol in Europe. When buying ISDN equipment, you need to know to what type of switch you will be connecting so you can make sure that your ISDN device supports that switch protocol.

ISDN B-CHANNEL PROTOCOLS

The B-channel protocol is the language that is spoken over each 64 Kbps B-channel between two ISDN devices that are connected. The main purpose for B-channel protocols is to transfer data across the B-channel in a standardized fashion.

READER'S NOTES

You can easily get into a situation where your ISDN Internet access device can speak with your ISDN switch but it cannot speak with the ISDN Internet access device on the other end of an ISDN call. In this case, your ISDN device supports the correct ISDN switch protocol, but it does not support the correct B-channel protocol.

Many B-channel protocols are defined in the world. Table 2.1 gives a list of some of the most popular protocols.

Table 2.1 A list of popular B-channel protocols.

V.110	Used to connect slower, pre-ISDN, communications devices to high-speed ISDN lines (primarily used in Europe)
V.120	Protocol used for similar purposes as V.110, but mainly used in the United States.
Point-to-Point Protocol (PPP)	Protocol that allows for carrying multiple types of LAN protocols over communication links (for example, Internet, Novell, Appletalk, etc.)
Multilink PPP (ML-PPP)	Extension to PPP that allows the aggregation of multiple B-channels for higher throughput

Table 2.1 Continued

BONDING	Bandwidth On Demand Interoperability Group. Another protocol for the aggregation of ISDN channels for higher throughput (used mainly for non-Internet applications such as videoconferencing)

The three most popular protocols for Internet access are V.120, PPP, and ML-PPP. The V.120 and PPP protocols will give you a 64 Kbps connection to the Internet. ML-PPP will give you a 128 Kbps connection to the Internet. If you are using one of these protocols, you will need to make sure that your Internet provider supports them as well.

INTERNET PROTOCOLS

While the purpose of this chapter is mainly to introduce the ISDN technology, it is important to discuss the relationship between the ISDN related protocols that run on your ISDN device and the Internet protocols that will run on your computer.

As you may already know, the main protocol for the Internet is called Transmission Control Protocol/Internet Protocol or TCP/IP. In dial-up access to the Internet, TCP/IP information needs to get from the computer, where it originates, down to the ISDN device, where it can be transmitted to the Internet service provider. On its trip from the computer to the ISDN device, this information may undergo several translations that will prepare it for transmission on the ISDN B-channel. In the modem world, this is usually done by taking the TCP/IP information, sending it to PPP or SLIP (Serial Line Internet Protocol), which runs on the PC, then sending to a modem for transmission. In the ISDN world, there are four principle architectures (illustrated in Figure 2.4) that may be used depending on the hardware and software involved:

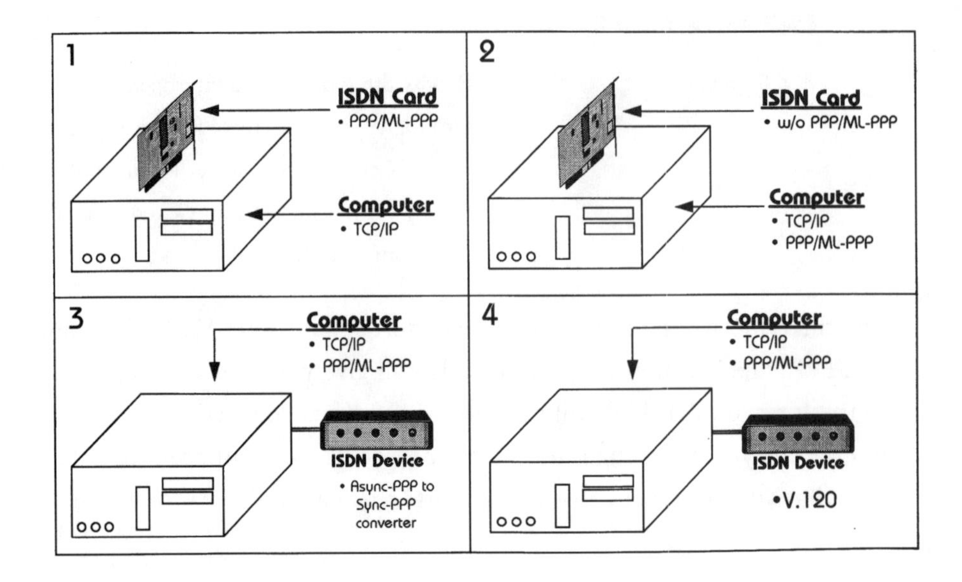

Figure 2.4 Internet-ISDN protocols relationship.

1. TCP/IP runs on computer, gets sent to PPP/ML-PPP, which runs on an internal ISDN computer card.
2. TCP/IP runs on computer, gets sent to PPP/ML-PPP, which runs on computer, gets sent to internal ISDN computer card, which passes the data through without any additional protocol.
3. TCP/IP runs on computer, gets sent to PPP/ML-PPP, which runs on computer, gets sent to communications port (RS-232 port), gets sent to internal or external ISDN device that converts the computer version of PPP to the ISDN version of PPP (this conversion is known as async-to-sync PPP/ML-PPP conversion).
4. TCP/IP runs on computer, gets sent to PPP/ML-PPP, which runs on computer, gets sent to communications port (RS-232 port), gets sent to V.120, which runs on internal or external ISDN device.

ISDN DEVICES

The basis for ISDN equipment use is described in a document published by an international committee that works to provide ISDN standards. The diagram, found in Figure 2.5 (and explored in more detail in Chapter 7), shows the types of equipment defined for use within BRI (PRI makes use of a subset of this).

In North America, the transmission line is considered to end at the U reference point. This is the point at which the typical 2-wire line comes into a building from the telephone company. The S (or S/T) reference point describes the 8-wire line that goes into most ISDN equipment. The network termination function lies between these two points.

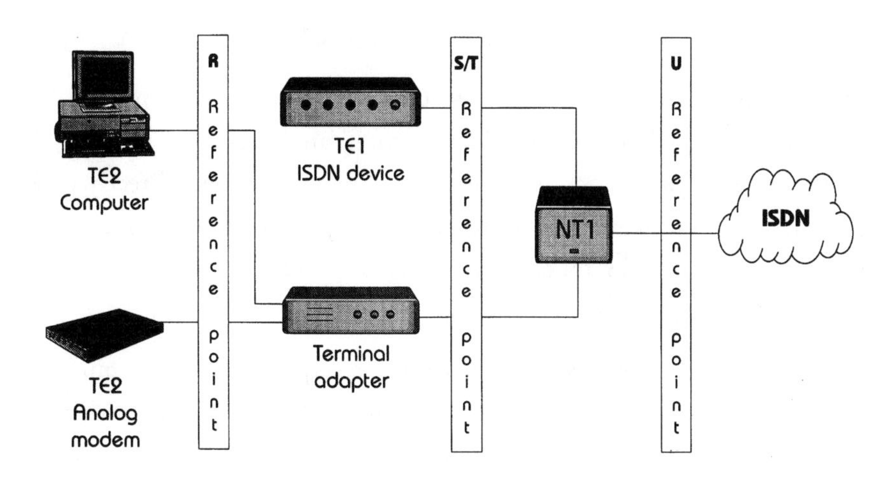

Figure 2.5 ISDN devices at reference points.

The Network Termination-1 function is needed to convert from the 2-wire line that is commonly used for analog equipment to the 8-wire line that is used by ISDN equipment.

READER'S NOTES

While the ISDN standards define an NT-1 as a device that performs a 2-wire to 4-wire conversion, in many cases, the S/T interface uses more than 4 wires and thus requires an 8-wire cable. Therefore, this book consistently refers to the NT-1 conversion as a 2-wire to 8-wire conversion.

Outside of North America, this conversion is considered to be part of the service provided by the local communications company—and, thus, an 8-wire interface jack will be installed for connection to ISDN equipment. Therefore, it is important for users of equipment outside North America to be sure that their ISDN device connects at the S/T point.

Awareness of this is also important to the user of ISDN equipment in North America. If the ISDN device provides an integrated NT-1, no extra NT-1 box needs to be purchased, and the 2-wire line will connect directly to your ISDN device. Otherwise, a separate NT-1 box must be bought, increasing the total cost of your ISDN equipment. This is discussed in greater detail in Chapters 4 and 5.

The diagram also mentions two other types of equipment—a terminal adapter (TA) and an ISDN terminal equipment (TE1). A TA provides an adaptation from one protocol to another. In this case, it provides a mechanism to use old pre-ISDN equipment on ISDN. In Figure 2.4, examples are shown of internal (examples 1 and 2) and external (examples 3 and 4) TAs. Examples of TE1 include LAN cards and other ISDN-specific devices that do not require interface conversion.

While this information may seem complex and unnecessary at this point in the book, it will become more and more pertinent when you proceed to buy and install the software and hardware that you will use for your ISDN connection to the Internet.

56 KBPS DATA CALLS VS. 64 KBPS DATA CALLS

When ISDN was created, the base data rate was established at 64 Kbps. Unfortunately for some U.S. ISDN subscribers, many of the local telephone, and long distance, companies can support only 56 Kbps out of the available 64 Kbps because of limitations of the telephone network. Until all telephone, and long distance, companies fully upgrade their network facilities to support 64 Kbps data calls, you may have to place 56 Kbps to make your call go through. Most ISDN devices support the selection of 64 Kbps or 56 Kbps calls (some equipment may do this for you automatically).

READER'S NOTES

56 Kbps calls are much more prevalent in long distance calls than they are in local settings.

UNLOCKING THE FULL SPEED POTENTIAL OF YOUR ISDN LINE

Until recently, most users of ISDN have only been able to use a maximum of 64 Kbps or 56 Kbps for their data connections. With the advent of protocols such as Multilink Point-to-Point Protocol, users can now aggregate their two ISDN B-channels to get a full 128 Kbps out of their ISDN line. If you want to get a 128 Kbps connection to the Internet, ML-PPP will most likely be the protocol that you use.

While ML-PPP is destined to be the most standardized 128 Kbps protocol for Internet access, some Internet providers support proprietary protocols that can also combine multiple B-channels. In these situations, the Internet provider should inform you of the type of equipment you need to connect to their service. Be aware that, if you choose to go with a product that uses a proprietary protocol, your interoperability with services other than Internet (or even with other ISPs) will be limited. It is always best to go with a standardized protocol such as ML-PPP or V.120.

Another route to getting extra speed out of your ISDN line is to use compression technology. In the modem world, it is almost a given that your modem will support compression (that is, V.42bis). This is not yet the case with ISDN. Some products support compression; others don't. Using compression, an ISDN user can get, on average, around 2:1 compression. This means that, on a 64 Kbps link, you can realize an effective data throughput of 128 Kbps and, on a ML-PPP 128 Kbps link, you can realize as high as 256 Kbps. Remember that both sides of the connection will need to support compression if you want to get the extra speed.

If 128 Kbps or 256 Kbps is still not enough to satisfy your hunger for speed, you could consider buying a second ISDN line, and combine the four available 64 Kbps B-channels into a monster 256 Kbps ML-PPP connection capable of handling 512 Kbps of traffic with compression. While this is totally feasible, you may not find this capability at your Internet service provider for a couple of years.

SUMMARY

The information in this chapter is presented to give you a feel for some of the basic elements of ISDN and how it works. You should now understand that a large portion of getting your ISDN connection to work properly is making sure that your ISDN device is speaking the correct protocols to both the ISDN switch and the device you are calling. For instance, your ISDN device must speak the correct switch protocol to communicate with the local ISDN telephone switch over the D-channel, and your ISDN device and the ISDN device that you are calling must support the same B-channel protocol (for example, V.120, PPP/ML-PPP) to communicate properly. Subsequent chapters will build on this foundation to solidify your understanding of what it means to get a high-speed connection to the Internet via ISDN. For those interested in learning a lot more about how ISDN works, and where it came from, the second part of the book presents a more comprehensive discussion.

CHAPTER

3

ISDN TELEPHONE AND INTERNET SERVICE

TWO MAJOR CATEGORIES OF SERVICE

Two major interrelated components—the ISDN telephone connection and an ISP—must come together seamlessly to deliver acceptable ISDN Internet performance. They are illustrated in Figure 3.1.

On one side you need a properly installed and configured ISDN telephone connection from your Local Exchange Carrier (LEC) (shown as 1 in Figure 3.1); on the other side you need a reliable Internet service provider (ISP) that supports ISDN (shown as 2). If they are combined properly you will then be on your way to gaining the full performance of ISDN. This chapter will discuss both the telephone and Internet service and offer background information to assist you through the ordering process. In addition, it evaluates the options and alternatives that you may face as you work

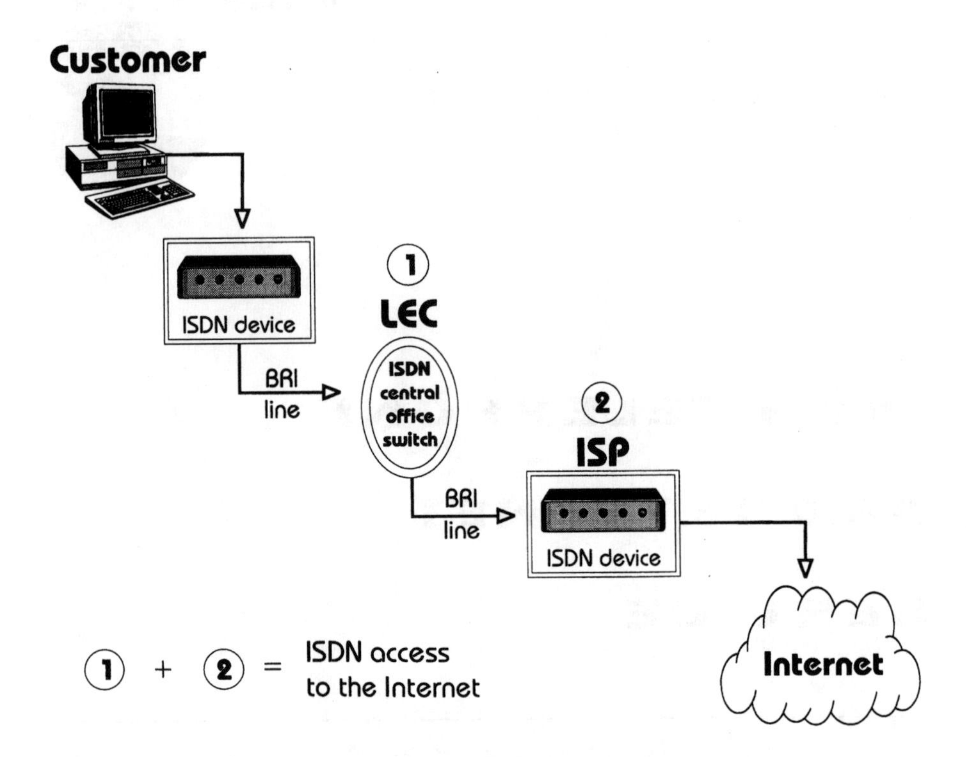

Figure 3.1 Two major components of ISDN service.

through the selection process to achieve ISDN access to the Internet. Chapter 3 presents the following topics:

- General availability of ISDN telephone service
- Determining availability of ISDN telephone service at your residence or office
- General Availability of ISDN Internet service
- Types of ISDN Internet service
- Locating and selecting an Internet service provider
- Ordering the ISDN Telephone Service

GENERAL AVAILABILITY OF ISDN TELEPHONE SERVICE

If we view the United States as Bellcore (Bell Communications Research, Inc.) does, illustrating the geography of the Regional Bell Operating companies (see Figure 3.2), we see seven major regions. (Other local and long distance companies offer ISDN services, but they are not discussed here because the RBOCs currently dominate most of the market.) It should be noted that Pacific Bell is a part of the Pacific Telesis Group.

By comparing the relative ISDN deployments of each of the RBOCs, we can begin to get a feel for the general availability of ISDN across the country. For those of you not serviced by one of the seven RBOCs, but by a different carrier, you will find the information presented here generally applicable.

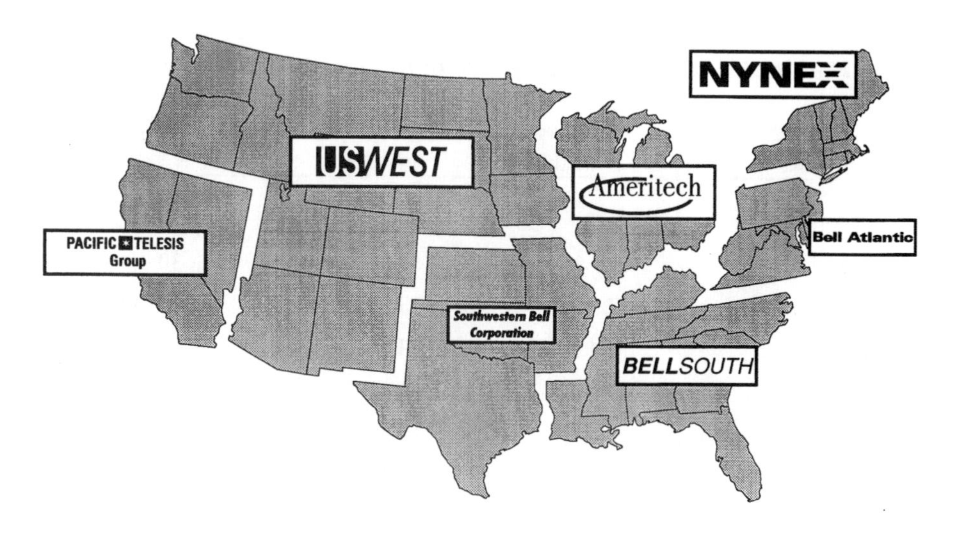

Figure 3.2 Seven major regional bell operating companies.

ISDN LINES VS. ISDN SWITCHES

To understand the availability of ISDN telephone service we will briefly discuss deployments of ISDN-capable telephone lines and ISDN-capable switches. This information, furnished in Table 3.1, is derived from the independent carriers and summarized periodically by Bellcore. This summary is current as of October 1994. Should you require more recent information, contact Bellcore or your local exchange carrier.

Within the seven regions of the RBOCs there are more than 120 million individual residential and business telephone lines, including every type of analog and digital service. On average, the growth from year to year (1995 to 1996, for example) in new telephone lines is expected to be approximately 10 percent. Of the approximate 120 million lines nationwide, according to the information in Table 3.1, about 88 million are capable of supporting ISDN service. Note that an *ISDN-capable* line is by no means an *ISDN-active* line. Becoming active requires a paying customer. The actual number of active ISDN lines is somewhat elusive at this time because the widespread availability of ISDN service is less than three to five years old, and residential service

Table 3.1 Deployment of RBOCs. ISDN Assets (CY1996).

RBOC	Regional Availability	# of ISDN Switches	# of ISDN Lines*
Ameritech	82%	618	15,124,000
Bell Atlantic	82%	624	16,630,000
Bell South	72%	734	14,900,000
NYNEX	53%	233	8,970,000
Pacific Bell**	86%	418	13,700,000
Southwestern Bell	76%	181	10,900,000
US West	53%	267	8,029,000

Source: Bellcore

* # of ISDN Individual lines that can be supported by the operating ISDN switches.
** Pacific Bell is a part of Pacific Telesis Group

is relatively new. In any case, the number of sold residential and business ISDN lines could be 1 to 2 percent of capacity. Even though these data indicate that the lion's share of all currently deployed telephone lines can support ISDN service, many believe otherwise. Unfortunately, in spite of ISDN's widespread deployment there are still isolated pockets of limited telephone service, especially in rural areas. When you contact your local exchange carrier for service, you may find that ISDN is not yet available in your area.

TYPES OF ISDN SWITCHES

In most regions of the United States the RBOCs have chosen to deploy three principal central office telephone switches that support ISDN telephone service. They are the AT&T 5ESS, Northern Telecom DMS100, and the Siemens Stromberg-Carlson EWSD. What is important to know about these switches is that they conform to the national standard (National ISDN-1); as such, they will be compatible with the ISDN equipment you will connect to your computer.

The numbers of each type of switch are not readily available within each region or local area. Therefore, Figures 3.3, 3.4, and 3.5 are shown only to establish the trends over a three-year period.

As is the case of ISDN-capable lines, ISDN-capable switches do not necessarily correlate to which company currently has the most or fewest active customers. This perspective is complicated by the size of the area covered by each carrier, population densities, and concentration of service in large cities. Nevertheless, the number of ISDN-capable switches is a measure of the carrier's capability to support the rapid growth in ISDN.

The RBOCs have invested heavily in building and deploying the infrastructure to make ISDN service available nationwide and to modernize their networks to support present and future requirements for data services. Nationwide you should be able to find ISDN available in about 70 percent of the country; by the year 2000, deployments should reach the 90 to 95 percent bracket.

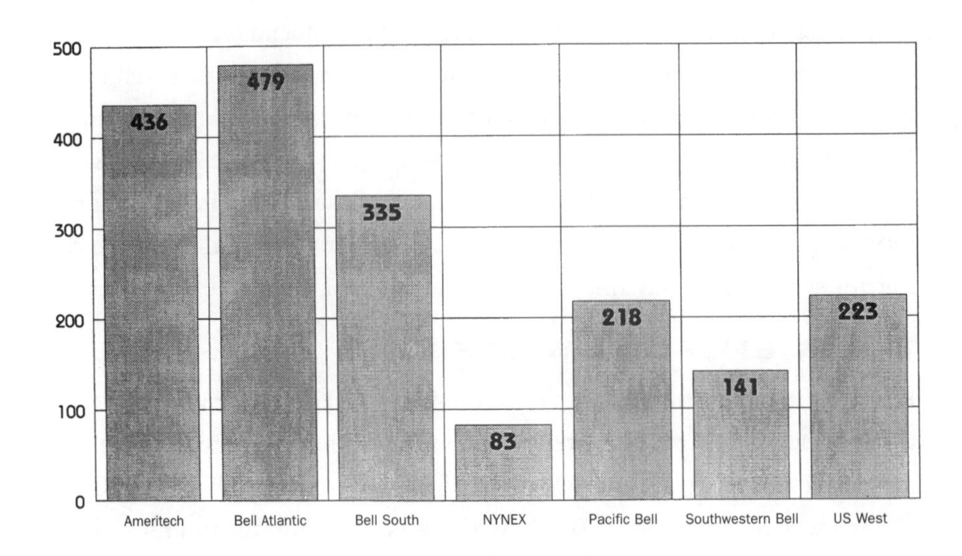

Figure 3.3 1993 ISDN switches deployed.

Source: Bellcore

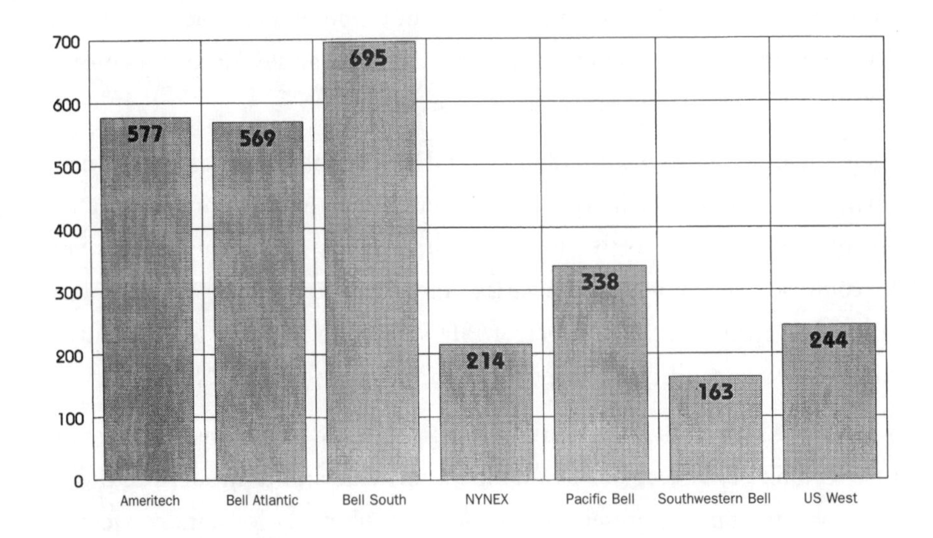

Figure 3.4 1994 ISDN switches deployed.

Source: Bellcore

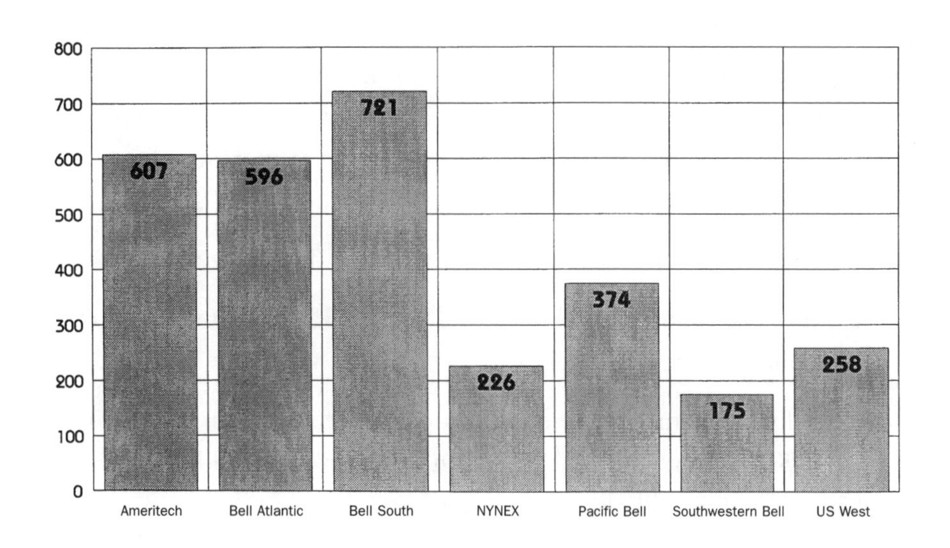

Figure 3.5 1995 ISDN switches deployed.

Source: Bellcore

DETERMINING THE AVAILABILITY OF ISDN TELEPHONE SERVICE AT YOUR RESIDENCE OR OFFICE

To determine if you can get ISDN service you need to contact your telephone company. The ISDN Information Hotlines for the RBOCs are as follows:

Ameritech	1-800-832-6328
Bell Atlantic	1-800-570-4736
Bell South	1-800-428-4736
NYNEX	1-800-438-4736
Pacific Bell	1-800-472-4736 (ISDN = 4736)
Southwestern Bell	1-800-992-4736
US West	1-800-246-5226

When you contact the telephone company, you will be asked to provide your phone number and address. This information alone could be sufficient to determine the availability of ISDN service and its charges. There could be several steps to the process of determining the availability of your ISDN service as follows:

1. Determine if your local central office has an ISDN switch.
2. If not, determine if another central office with an ISDN switch can support your service.
3. Finally, perform a "loop qualification" to see if the distance between the ISDN switch and your location is within specified limits. The quality of the line will also be determined.

In the situation when your local central office cannot support ISDN, the telephone company may be able to support the service from a central office outside your immediate dialing area. This is sometimes referred to a foreign central office (FCO) or foreign exchange (FX), as illustrated in Figure 3.6.

In these situations, Bell Atlantic, Ameritech, Bell South, NYNEX, and Pacific Bell have instituted roughly equivalent programs that will extend ISDN service to the customer at no additional cost. In some regions, though, you have to pay an additional charge to extend ISDN service.

For most individuals the ISDN information referral numbers listed earlier will be sufficient to determine the availability of ISDN service. If, however, you would like to research the availability of ISDN service prior to contacting your local exchange carrier, and if you have a connection to the Internet, several options are available on-line. You can use an on-line form set up by the LECs or several hardware vendors as a value-added service. Appendix A provides a list of Uniform Resource Locators (URL) that will help you obtain this information. For most inquiries all you have to do is enter your area code and 3 digit prefix.

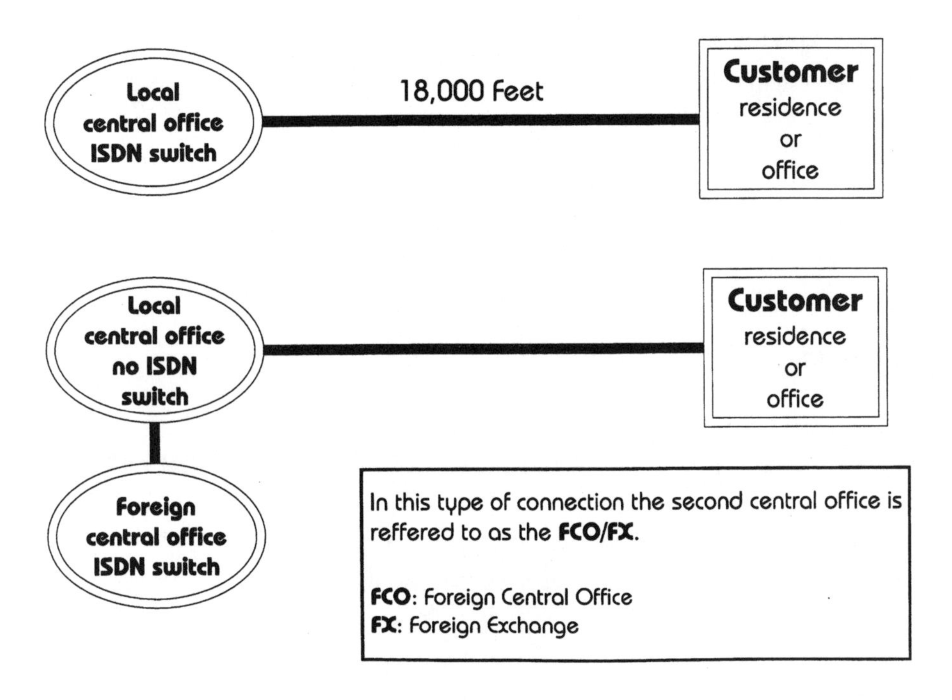

Figure 3.6 Providing ISDN service.

LOOP QUALIFICATION

ISDN, like many other forms of telecommunication, has a theoretical distance limit beyond which the transmission requires some form of reinforcement. For ISDN communications within a specified region, the theoretical maximum distance between the switch/central office and the customer is approximately 18,000 feet.

Loop qualification involves checking the physical distance and quality of the line between your residence or office and the local central office. If the maximum distance is exceeded, special equipment, such as a signal amplifier or a repeater to strengthen the signal, may be required. In most regions a technician actually will have to connect the test instrument to the specific line in question. However, in some regions (for example, Bell Atlantic's region), specific software can automate the process. When this software is used, you generally will know in a very short time if the line qualifies.

READER'S NOTES

Loop qualification is a standard procedure conducted by the telephone company to determine if the service can be provided to your facility. The test is usually completed at no additional cost to the customer.

TYPES OF ISDN TELEPHONE SERVICE

Local exchange carriers have come up with a wide variety of labels and names to market their ISDN services. Many of these names boil down to two generic types of service as shown in Table 3.2.

The difference between a BRI line for a residence or business is mainly the cost and feature set of each. The business line usually costs more and has expanded features.

READER'S NOTES

PRI service is used only for large business requirements; it is very expensive. It is only mentioned here for completeness.

Table 3.2 Generic types of ISDN service.

Type	Capacity
Residential	Basic Rate Interface (BRI): 128 Kbps
Business	Basic Rate Interface (BRI): 128 Kbps
	Primary Rate Interface (PRI): 1.544 Megabits

RESIDENTIAL SERVICE

Residential ISDN service and associated tariffs are not uniformly available across the country. Given the lack of infrastructure to support a widespread residential customer base, a surge in residential ISDN service was not possible until recently. Pacific Bell has notably led in the introduction of this service; it seems to be driven largely by the demand for telecommuting requirements in the San Francisco area. Bell Atlantic is offering a residential service throughout its region, with impressive targets for full deployment of the service. Other telephone companies will follow as demand for high-speed access to the Internet and telecommuting grows in the respective regions.

In the final analysis, customers not only will be looking for the performance of ISDN, but also will be influenced by the cost of the service. Usage rates (per minute charges) for data calls, billed against each B-channel in use, have been a major detractor for the residential service. ISDN advocacy groups are pressing for rates comparable with those for standard analog service, but this kind of parity may not be possible for some time.

BUSINESS SERVICE

ISDN has been sold as a business service since its inception. Its advanced voice and data features have made the technology an attractive alternative to Private Branch Exchange (PBX) telephone systems. Today, most ISDN lines deployed in the United States are serving businesses. Large corporations are using both the BRI and PRI services while small businesses and branch offices are using the BRI service. ISDN business service is available in major urban centers across the country.

GENERAL AVAILABILITY OF ISDN INTERNET SERVICE

Internet access via ISDN is available at national, regional, and local levels. However, the number of ISDN Internet service providers is growing so rapidly that it is difficult to say how many there are and where they are located. As ISPs discover the performance of ISDN, more and more are making this solution available to their subscribers.

To add an element of possible confusion in this highly competitive and complex market, telephone companies, both local and long distance, are entering the Internet access business. RBOCs such as Pacific Bell and Ameritech already offer ISDN Internet service and many others are planning to offer the service. The long distance carriers, such as MCI, Sprint, and now AT&T, who offer Internet access, could also develop ISDN access solutions.

NATIONAL ISPS

In the United States today there are relatively few choices at the national level. A national provider is defined here as one that has local access points distributed across the country, commonly referred to as "Points of Presence" or POPs. See Figure 3.7.

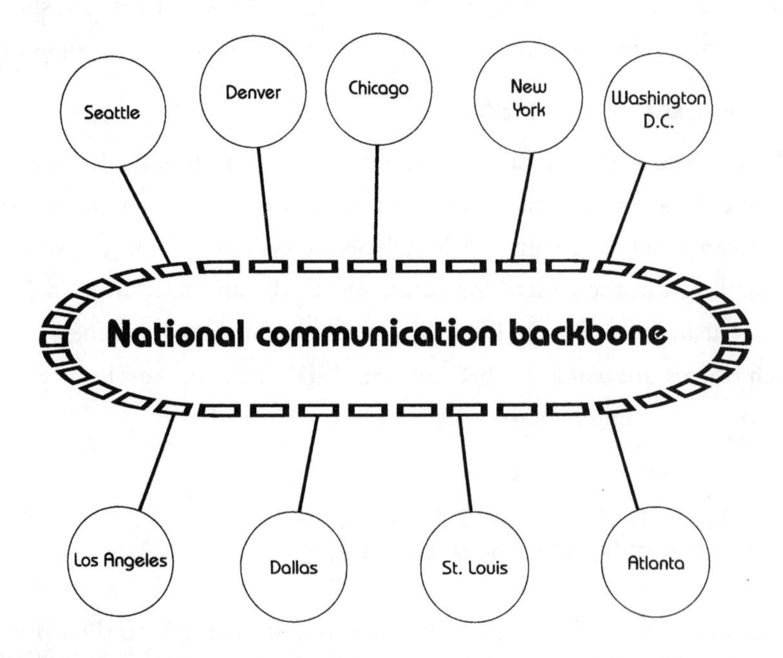

Figure 3.7 National Internet service providers' points of presence.

One of the most experienced national ISPs is Performance Systems International, or PSI Net. PSI Net has offered both dial-up LAN and individual access at 64 Kbps for several years and they now support 128 Kbps. Having recently gone public, PSI's network is expanding to offer service in more than 200 locations around the United States. The costs and features of PSI ISDN service are discussed later in this chapter.

A number of benefits are associated with using a national provider. One of the best reasons is that a national provider's services will be standardized across the country. A large company with offices in different cities would be able to use the same equipment and secure the service at multiple locations with a single contract. The main disadvantage is that national service providers have limited choices of services and, in some situations, cannot adapt quickly to changes in the technology because of the large investments required to modify the infrastructure.

LOCAL AND REGIONAL ISPS

Local and regional ISPs are also providing ISDN connections at an accelerating pace. More than 100 ISPs advertise ISDN service across the country, and more are being added almost every day. At the moment, the largest number of ISDN offerings from local providers seem to come from California, Texas, and Washington D.C.

READER'S NOTES

Appendix E provides a partial listing of ISDN-capable ISPs across the United States.

Wonder what the difference is between a national ISP and a local ISP? The answer is probably very little in terms of the type of ISDN service you are likely to be offered. However, the small local provider may be more willing to tailor its service to a specialized requirement, or it may be in a better position to adapt more quickly to new solutions. Your choice of an ISP, in the end, will more than likely be based on demonstrated performance, cost, and good technical support.

OTHER TYPES OF INTERNET PROVIDERS

On-line services, including CompuServe, Prodigy, and America Online, are also developing ISDN access solutions to introduce across their national networks. Most of these services already offer their customers access to the Internet. ISDN access is being added to address demands for higher speed.

One of the latest entrants on a national scale is the Microsoft Network, which offers Internet access through Windows 95. Microsoft has teamed with UUNET, one of the first commercial Internet service providers, to develop and operate a national, and perhaps global, network. UUNET currently offers ISDN service.

1-800/LONG DISTANCE SERVICE

In a situation where you may have ISDN telephone service but no local ISP, your options are obviously limited. In this case, you will have to connect either through a 1-800-ISDN ISP such as CERFNET, based in California, or incur the long distance charges to the nearest ISDN-capable ISP.

TYPES OF ISDN INTERNET SERVICE

There are really only two kinds of ISDN/Internet service available at this time, dial-up and dedicated service. Dial-up service refers to a temporary connection that only gets established when accessing the Internet. Dedicated service refers to a permanent connection that will generally stay up 24 hours a day. Large office LANs and operation of Web servers may require dedicated 24-hour connections to support the open availability of their service. Dedicated ISDN is sold in most regions as a configured service named Centrex.

Dial-up and dedicated services can be configured to support either 1B or 2B-channels, that is 64 or 128 Kbps service. The ISP will also charge accordingly. As far as the equipment goes, in some instances you may be limited to a 64 Kbps solution by the hardware selection; in other cases, even if the hardware is capable of 128 Kbps bandwidth, your Internet provider will allow you to connect only at 64 Kbps.

Typical dial-up and dedicated networks are shown in Figure 3.8 and Figure 3.9.

For dial-up service, your computer(or LAN) is connected to an ISDN device that is plugged into the telephone jack. At the Internet service provider your packets are received by a similar ISDN device and then routed over the Internet. As the name indicates, the user must *dial up* the Internet service provider to begin an active session. At the completion of

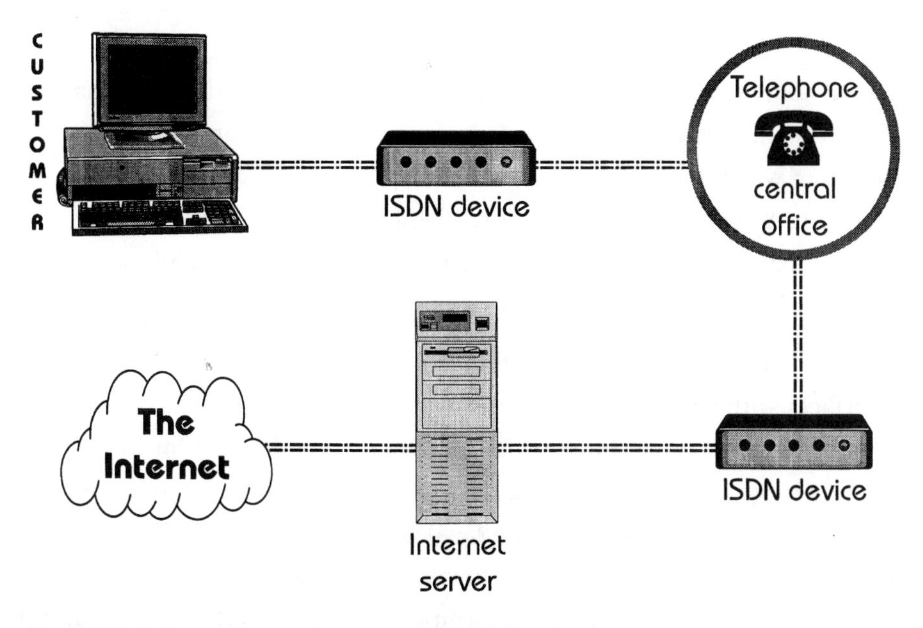

Figure 3.8 64/128 Kbps ISDN dial-up network.

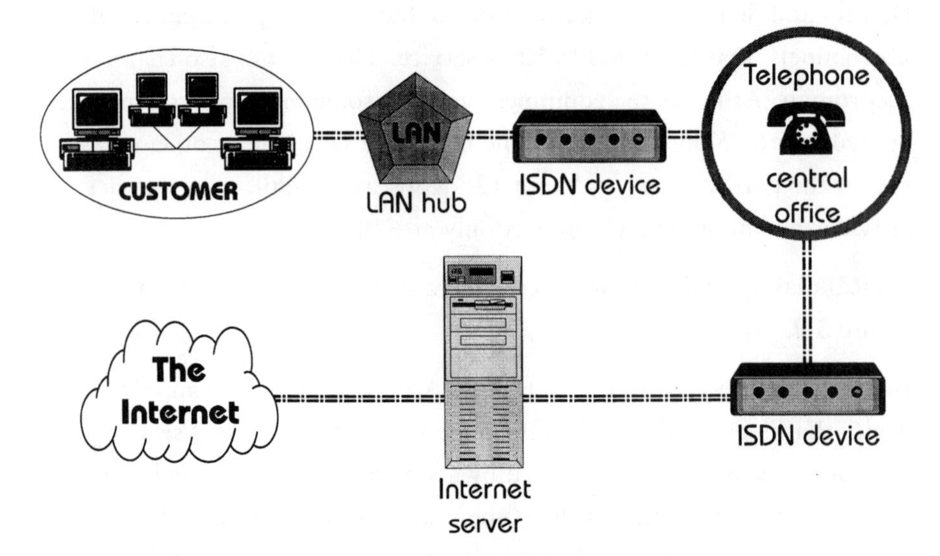

Figure 3.9 128 Kbps ISDN dedicated network.

the session, the telephone connection is broken and the call is terminated. This way the user pays only telephone usage charges, if applicable, for that period in which the line is in use. The telephone usage charges can vary from 2 to 8 cents per minute for each B-channel. Even though the per minute charges seem small, if the line were to be left open inadvertently you could receive an extraordinary bill at the end of the month.

For the dedicated service, you will most likely be using an ISDN device that interfaces with your office LAN (instead of just one computer). The ISDN line stays connected 24 hours a day, and you pay a flat rate for the telephone service. Regardless of what type of ISDN device selected, the same equipment will generally be recommended for use on both sides of the connection. This is usually done to optimize end-to-end performance and to minimize maintenance issues. How you connect to the ISDN service and the types of equipment you will require will be discussed in Chapters 4 and 5.

READER'S NOTES

Most ISPs do not offer many choices regarding how you can subscribe to their service. ISPs like to find solutions that work with their individual network architecture and then require their customers to use that solution. In some ways this approach limits your choices but, from another perspective, at least you know that the solution most likely will work.

SELECTING AN INTERNET SERVICE PROVIDER

Selecting an Internet service provider has been addressed in numerous books, magazines, and on-line newsgroups. For the most part, it is like selecting any other type of service needed to satisfy an individual or business need. The selection criteria in these situations—performance, price, and responsive technical support—apply as well for ISDN service.

CHOOSING THE SERVICE THAT SUITS YOUR NEEDS

Before selecting the ISP, you should spend some time determining which applications will be most important to you. For example, will your main applications require text-only access, or will you be downloading large graphic files from the World Wide Web? You should also consider the amount of time you expect to use the service each day, the capacity of the computer platform you will use to access the service, and, last but by no means least, how much bandwidth you will need for typical applications. These considerations won't necessarily lead you to an ideal or optimum solution, but they could raise some issues in your selection process.

HOW TO DETERMINE THE AMOUNT OF BANDWIDTH REQUIRED

Avid Internet users will almost inevitably desire all the bandwidth they can afford. However, because bandwidth is priced according to the amount you use you may be able to save some money if less will do the job. Conversely, if less won't do the job, knowing what you need—and what that will cost—before you sign up for the service is advantageous.

One way to get a feel for how much bandwidth you might require is to determine the amount of time it takes to download files of different sizes from the Internet. Once you know the download time, you can then make an educated guess about the amount of bandwidth you require. Figure 3.10 provides the time it will take to download three different size files at four different ISDN transmission speeds.

Now that you are armed with two essential pieces of information—ISDN service is available at your home or office and your application requires 64 Kbps or 128 Kbps of bandwidth—you are ready to face the task of shopping and selecting an Internet service provider.

LOCATING THE INTERNET SERVICE PROVIDER

Locating an ISP offering ISDN service should not be difficult. Finding the right one for your particular needs and budget may not be as easy. When

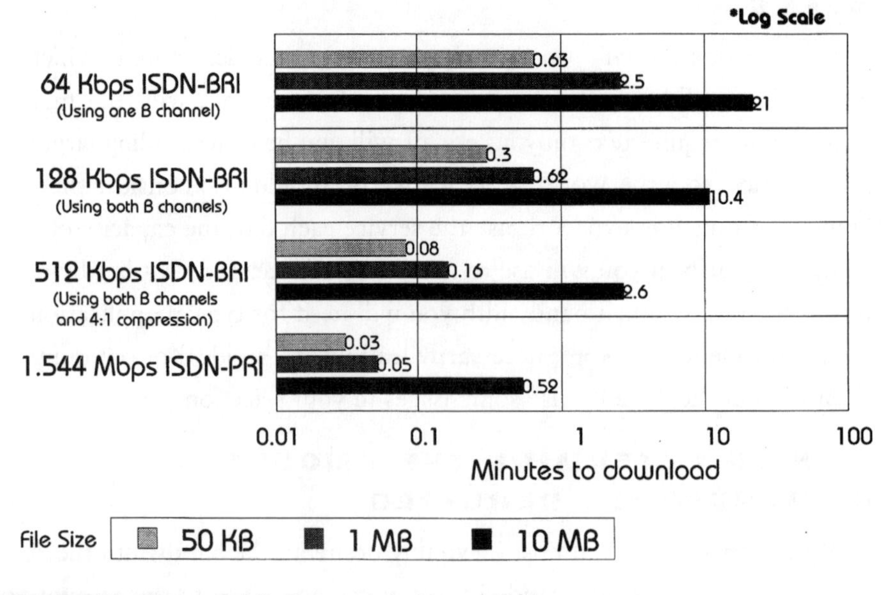

Figure 3.10 Determining bandwidth.

READER'S NOTES

Select the bandwidth/transmission speed that readily supports your main or most frequently used application. Experienced surfers on the Internet know that the speed at which your equipment can access the Internet does not always match the speed at which the Internet responds. The ultimate speed you realize will be affected by the current amount of traffic on the Internet plus the size of the communication link and type of equipment existing at the location from which the data is being accessed.

you buy a car, in most cases, you have a chance to drive it first to see if it meets your expectations. If it does, you buy it. Because most computers now have analog modems, you may already be equipped to check out an Internet provider's analog dial-up service assuming that the ISP offers a trial option to new customers.

This trial approach, unfortunately, does not extend readily to most new ISDN users for several reasons: Most new ISDN users do not have the proper device available to test the service, and the proper ISDN device should not be purchased until all the other information requirements have been satisfied regarding your service. Trying out an ISDN service is, therefore, more complicated.

You have three alternatives:

1. Use blind faith: go with the first ISP that sounds good.
2. Follow friendly advice: talk to your friends and associates, find a satisfied customer, and go on his or her recommendations.
3. Analyze options: First, talk to several ISPs and seek an opportunity to speak to one of their customers; then, see their services at their own facilities or at one of their customer's; last, attempt to get answers to questions that will assist you in making an informed decision.

LIST OF QUESTIONS FOR ISPS

When screening ISPs, you should ask the following basic questions:

- Does the ISP support 64 Kbps?
- Does the ISP support 128 Kbps?
- Does the ISP support 128 Kbps compressed?
- What is the type and capacity of its connection to the Internet?
- What type of equipment does the provider recommend?
- Can you see the performance at a demonstration facility?
- Can the ISP furnish ISDN customer references?

TWO NATIONAL ISDN INTERNET SERVICE OFFERINGS

As we have discussed earlier, ISDN Internet access really boils down to a matter of bandwidth. Service providers have been very creative in packaging the content, hardware, and pricing of their service, but, in every case, you will be comparing a dial-up or dedicated account at 64 Kbps or 128 Kbps.

Appendix E lists a representative sample of ISPs across the United States offering ISDN service. If you can't find several in your area to consider, refer to Appendix A, which lists information sources on the Internet that will lead you to more complete and more up-to-date sources.

To provide some insight on how ISDN service offerings are configured, PSI Net and UUNET ISDN service offerings are described.

PSI NET SINGLE AND LAN-ISDN SERVICE

PSI Net as one of the largest Internet providers in the United States offers numerous Points of Presence for that single PC dial-up account as well as

LAN connections. The service supporting the single PC user is known as InterRamp for dial-up customers. PSI Net provides a complete solution which includes all of the necessary software and hardware.

PSI Net's service for connecting up a LAN is called LAN-ISDN. Some features of the LAN-ISDN service are:

- "On demand" Internet ISDN access
- LAN access to the Internet
- ISDN 128 Kbps plus TCP/IP internetworking
- Nationwide local access
- USENET news feed included

PSI has upgraded their service to allow the customer to use an entire GRI digital circuit for Internet access at up to 128 Kbps. PSI's LAN-ISDN service may also be enhanced in the future to support 512 Kbps service through standard compression features.

PSI Net has priced their InterRamp service at $29 per month for 29 hours of service plus $1.50 per hour for each additional hour during prime time. They also charge a nominal set-up fee of $9 and $50 for the supporting software. PSI Net also provides an additional incentive which gives free access from 11 p.m. to 8 a.m. 5 days a week and all weekend. InterRamp service is projected to be offered in 225 cities by the end of 1995.

UUNET'S ISDN WORKGROUP INTERNET SERVICE

UUNET's service is configured to provide business users a high-speed dial-up connection to the Internet. Users who are part of a local area network can access the following features:

- E-mail
- USENET news groups

- Log-on to remote computers (telnet applications)
- Location and retrieval of files from the Internet (Archie and FTP)
- World Wide Web browsing

The service can be purchased with a single B-channel delivering 64 Kbps or 128 Kbps over two B-channels. The service charges (Table 3.3) are based on the throughput provided and can be as low as $95 per month for a single B-channel, based on a 24-month commitment. Includes 50 hrs/mo of service. Additional hours are $2.00/hr. Start up charges will also be incurred with this service.

UUNET recommends that the customer use the Ascend Pipeline 50 ISDN router, which can be purchased on the open market or from UUNET, at the customer site. In many areas of the country there is a per-minute usage charge for each B-channel when data are being transmitted. This additional charge should be factored into your overall cost equation as you compare service offerings.

In addition to business LAN connections UUNET has also released a low-cost ISDN service for individual residential users.

ORDERING THE ISDN TELEPHONE SERVICE

Table 3.3 ISDN UUNET's workgroup pricing for 64 Kbps or 128 Kbps Internet service.

Cost includes all standard features, plus unlimited connect time

	1 B-Channel	2 B-Channel
Startup charge	$295.00	$395.00
Standard monthly charge (50 hours of service)	$ 95.00	$195.00

If you have been able to work your way through each of the previous steps, at this point you have done the following:

1. Determined that ISDN telephone service is available at your residence or office
2. Thought about your applications and requirements for bandwidth; are satisfied that your computer has sufficient speed and memory
3. Requested information and price quotations from several ISPs
4. Tested or witnessed the service performance
5. Decided on which service provider you want to use
6. Determined generally what hardware you will be purchasing

Equipped with the proper information from your ISP and the ISDN line-ordering information from the hardware manufacturer, you are ready to contact the telephone service provider and order the ISDN BRI line. The important considerations will be equipment compatibility, configuration options, and pricing.

RELATIONSHIP BETWEEN YOUR HARDWARE CAPABILITIES AND THE TELEPHONE SERVICE

As discussed in the last chapter, ISDN standards were set forth to make it easy for all the different types of ISDN equipment to interoperate. Unfortunately, more than 10 ISDN world standards now define the way an ISDN device will interface with the ISDN telephone network. When you order ISDN service you therefore need to know which standards are implemented on your local switch. The National ISDN initiative has helped this issue by defining one standard for North America for all switch developers (AT&T, NTI, and SSC) and ISDN customer premise equipment developers. Now you can be reasonably assured that if your switch is NI-1 based and your ISDN device is NI-1 based, both are interoperable.

There is still one major problem. When installing your ISDN line, the telephone company has to enter many parameters into the ISDN switch to make it work properly. Some parameters are important enough that if they are entered incorrectly, your ISDN device may not be able to work properly. To combat this problem, most ISDN vendors include with their product a list of ordering information that defines the necessary parameters for each of the major switch types. You can use this information to order the line.

In addition, a new procedure has been instituted to simplify the ISDN ordering process. This procedure uses a set of "standard ISDN ordering codes" to limit the permutations of switch parameters that must be entered to make a given product work. These ordering codes are based on 17 feature sets called capability packages that were developed by the National ISDN Users' Forum (NIUF). These capability packages are listed in Appendix C.

If a product supports these ISDN ordering codes and the telephone company is prepared to use them, you can simply give the ordering code for the product and the ISDN line, in theory, should be installed properly. Look for names like Acme Red or NUCO Blue as sample ordering code names.

ISDN LINE CONFIGURATION

There are potentially hundreds of parameters that must be programmed into an ISDN switch to configure an ISDN line. Fortunately, you will have to worry about only a very small subset of these parameters. The fundamental pieces of the line configuration are the switch type, number of data and/or voice channels, and number of service profile identifiers (SPIDs). The SPID is a 9–20 character numeric value that uniquely identifies an ISDN device on an ISDN line. Basically it is a numbered code that allows you to access the ISDN telephone network.

For an ISDN Internet user the basic configuration will be an 2B+D line with an NI-1 switch type, alternate voice and data on both channels, and two SPIDs. You will want an NI-1 line if possible because it is the most standardized interface available in the country. 2B+D will allow you to connect to the Internet over two simultaneous 64 Kbps calls. Alternate voice and data on both channels will enable you to support both data applications such as Internet access and voice applications such as an analog telephone or modem on the same telephone line. Two SPIDs may be necessary to support two simultaneous data calls on certain types of equipment. This configuration has proven itself with almost every type of ISDN Internet equipment around.

On top of this baseline configuration, you may want to add more features to the line if you are supporting voice applications on the line. These extra features are similar to those found on normal analog phones. Call hold, call transfer, conferencing, and additional call offering (also known as call waiting in the analog world) are a few of the most popular.

Please keep in mind that the more parameters added to the ISDN line, the more difficult it may be to get the line working properly. Unless certain types of extra features are absolutely essential, you should stick with the baseline features first. For this reason it is essential that you become familiar with the product and its associated documentation, which, in most cases, will include a script for ordering the telephone service.

It is useful, as a preventive measure, to give the telephone company a copy of this ordering information and to have it acknowledge its familiarity, or lack thereof, with the specific product that you want to use. Should you be unable to locate that information with the normal product manuals, you can call the vendor directly or, if you have access to the Internet, you can sometimes download the information from the vendor's home page. Appendixes A and B show contact information and a representative sampling of ISDN hardware specifications.

ISDN TARIFF SUMMARY

Tariffs for ISDN telephone service depend on several factors. As with all telephone services, charges for service are controlled by state Public Utility Commissions (PUCs). Rates can differ considerably within the area of a particular RBOC, depending on how it chooses to package the service and the actual cost to the company to deliver the service.

Telephone charges are, unfortunately, a complicated business. A cynical view of the situation might be that *you really won't know what it costs until the bill arrives,* which may be an oversimplification. Those responsible for providing a single source of information on charges for the RBOCs such as Bellcore, have difficulty maintaining an accurate, up-to-date consolidated listing of the many and varied approaches to determining charges across the country.

This section attempts to summarize the information for the seven major RBOCs for the ISDN-BRI business/residential service. Because tariffs change periodically you will need to contact the telephone company directly for the latest figures that apply to your specific location.

PRICING OF ISDN TELEPHONE SERVICE

The pricing of ISDN telephone service varies across the country. Each of the seven RBOCs appears to have a different set of formulas to calculate the price of its ISDN service. This complexity makes it very difficult to explain and even more difficult for the average person to understand. To gain approval for charging these prices each RBOC must go through a review and approval process before the Public Utility Commission(s) governing its area. The RBOC must present a case that justifies the charges to be established based on all the factors and costs of delivering that service to the customer.

The main components of ISDN telephone service pricing are setup or installation charges, monthly charges, and usage charges. Setup and installation involve checking whether you can receive the service, configuring the ISDN switch, possibly adding to the switch capacity, and finally coming to your residence or office and making the final connection to the designated telephone jack. Any type of wiring work that will need to be completed will be charged at local rates for time and material.

Table 3.4 Estimated ISDN pricing (2B+D alternate voice and data).

RBOC	Installation	Monthly Charge
Ameritech	$160	$40
Bell Atlantic	$200	$50
Bell South	$150	$90
NYNEX	$310	$35
Pacific Bell	$70	$30
Southwestern Bell	$485	$50
US West	$70	$35

Note: Pricing may vary considerably from state to state within each region.

Table 3.4 lists approximate costs for installation and monthly charges. A more complete discussion of the ISDN telephone services and pricing offered by the RBOCs, including line usage charges where applicable, is presented in Appendix D.

SUMMARY

Many questions need to be addressed in the selection process for ISDN telephone and Internet service. For telephone service, the questions mainly relate to the ISDN line configuration and the ISDN device interface parameters. Unfortunately, there is no way to evaluate the performance of the telephone company without going through the expense of getting connected. For the Internet service, however, you may be able to evaluate the service before you pay.

A few general questions are listed here as reminders. There is no substitute for a trial performance evaluation, however, if it can be arranged. Figure 3.11 provides an open form that may assist you in organizing and evaluating the service.

	Cost of Service			Comments
	One Time	Monthly	Usage	
Telephone Service				
Set/up Installation				
1B/64 Kbps				
2B+D/128 Kbps				
Centrex/24 Hour				
Miscellaneous				
Internet Service				
Dial-Up				
Set/up Installation				
1B/64 Kbps				
2B+D/128 Kbps				
Number of Hours/Day				
Charge for Additional Hours				
Hardware				
Software				
Miscellaneous				
Dedicated				
Centrex/24Hr/128Kbps				
Cost of Equipment				
Installation of Equipment				
Hardware				
Software				
Miscellaneous				
TOTAL:				

Figure 3.11 ISDN Internet evaluation form.

TELEPHONE

- Is ISDN service available at your location?
- What are the charges for the service: installation fee, monthly charge (including tax), any type of usage charges for the type of service provided?
- How long will it take to install the service?

- Be sure to ask the telephone company for all the line information including:
    - Order number
    - Type of ISDN switch (AT&T 5ESS, Northern Telecom DMS-100, Siemens Stromberg-Carlson EWSD)
    - Telephone number(s)
    - Service Profile Identifications SPID/SPIDs (one SPID or two SPIDs)

INTERNET

- Determine first whether the Internet service provider has experience in connecting customers using ISDN communications.
- Obtain some information on what categories of ISDN service are being offered. Does the ISP offer both dial-up and dedicated ISDN service?
- Inquire about how the Internet service provider network is supported. For dial-up service, how many ISDN lines are available for dial in? Will you be assured of an open line when you dial in for a connection?
- Determine whether there is a preference for any particular ISDN interface hardware. With some Internet service providers, they will permit the usage of only one particular type of ISDN equipment, which predetermines what you have to buy and generally what you have to pay for the unit(s). In some cases, though, the Internet service provider maintains an open network architecture and will consider using a variety of product solutions as long as they are shown to be compatible with the rest of their network.
- Understand how the service is priced.
- Request references of other customers to learn of their experiences and the technical support available when and if problems arise.

CHAPTER

4

CHOOSING YOUR HARDWARE AND SOFTWARE FOR ISDN INTERNET ACCESS

Once you have ordered your ISDN line from the telephone company and signed up with an ISDN Internet service provider, you need to think about the software and hardware you will need to purchase. Some ISPs will tell you exactly what software and hardware you need to interact with their service. Others will tell you what ISDN protocols (V.120, PPP, etc.) they support, and what ISDN speeds (64 Kbps, 128 Kbps) they support and leave it up to you to buy the software and hardware. In either case, it is important to get a feel for the different hardware and software solutions available to you. When upgrading your computer to support an ISDN connection to the Internet, you want to be prepared to make the correct choices.

GENERAL FACTS ABOUT ISDN HARDWARE

Before discussing the various types of hardware, let's go over some general points about ISDN hardware.

ISDN DEVICES REQUIRE POWER

ISDN devices are like any other communications devices in that they must be powered to be used. If you plan to use your ISDN line as a replacement for your current analog phone line, you may want to rethink that decision. If there is a power failure and you need to make a phone call, you are out of luck unless you own a battery backup for your ISDN equipment. It is always a good idea to keep at least one analog line for emergencies.

NO TWO DEVICES HAVE THE SAME FEATURES

Every ISDN product has slightly different features. This condition is in contrast to the modem world, where almost every modem has the same capabilities. In ISDN, you need to look long and hard at each hardware platform to decide if it has all the features you require for current and future requirements. Conversely, because you pay extra for each feature, there is no point in paying for features you do not need. When the final selection is made, there is likely to be some compromise. The following is a list of the most common ISDN device features:

- One or two analog ports (to support analog phones, modems, or fax machines)
- Compression
- Integrated NT-1 (network termination 1)
- Various ISDN protocols (V.120, PPP/ML-PPP, BONDING, etc.)
- U interface and/or S/T interface
- Channel aggregation (to support speeds greater than 64 Kbps)

RULES OF THUMB TO USE WHEN BUYING ISDN HARDWARE

When it comes to buying a piece of ISDN hardware, many different criteria can be used. A few criteria, however, should stand out in your decision process.

NATIONAL ISDN-1 SWITCH STANDARD

In North America, NI-1 is by far the most highly deployed switch standard. If you get an ISDN device that supports NI-1, you will have the greatest chance for interoperability today and in the future. The next best switch protocol for your ISDN device to support is AT&T Custom.

BEWARE OF PROPRIETARY B-CHANNEL PROTOCOLS

The widespread use of ISDN has been hindered for many years because of the lack of interoperability between ISDN devices. The interoperability between ISDN devices and ISDN switches has already been solved with the NI-1 standards. There are still interoperability problems between two ISDN products communicating over the 64 Kbps B-channel. Different vendors develop protocols that work only with their equipment, thus making interoperability impossible. Over the last year or so this has been changing as vendors implement interoperable standards such as V.120 and PPP/ML-PPP. When you buy a product that supports these open standards, you stand a better chance of being able to connect effectively with other ISDN devices from other vendors. For your Internet access to work smoothly, interoperability is a crucial requirement.

BUY PROVEN SOLUTIONS

One of the biggest problems with new ISDN products is that they take a long time to mature. It could take two or three software revisions before a product stabilizes. For this reason, it is suggested that you buy only proven ISDN solutions from proven ISDN companies.

ISDN DEVICES USED FOR INTERNET ACCESS

Four main types of ISDN devices are used to access the Internet: NT-1s, terminal adapters (TAs), ISDN LAN cards, and routers. This section introduces each category of devices, discusses how they work, and assesses some of the issues involved in using them.

NETWORK TERMINATION DEVICES—NT-1S

The Network Termination 1 or NT-1 is a device that converts the 2-wire ISDN line from the telephone company, called the U interface, into the 8-wire S/T interface (please recall from Chapter 2 that not all eight wires of the S/T interface are used). The U interface supports the connection of one device, and the S/T interfaces supports the connection of up to eight devices. An NT-1 is used, therefore, when a user wants to connect multiple ISDN devices to one ISDN line. For example, if you want to have an ISDN telephone on the same line as the ISDN Internet device, you will need an NT-1. Figure 4.1 illustrates this point. Even if you currently need to connect only one device, you may want to consider going the NT-1 route to have the flexibility to support additional devices in the future. Although the maximum number of devices that can be connected to an NT-1 is eight, only one or two devices are generally connected. Note that if you plan to use an NT-1, all of the other devices that you use for your ISDN line will have to support the S/T interface.

Some NT-1s support extra functionality on top of their main purpose of performing the U to S/T interface conversion. NT-1 Plus devices, as they are sometimes called, may have other capabilities such as analog phone ports and battery backup. NT-1s cost $300–400 and can usually be purchased from your local telephone company.

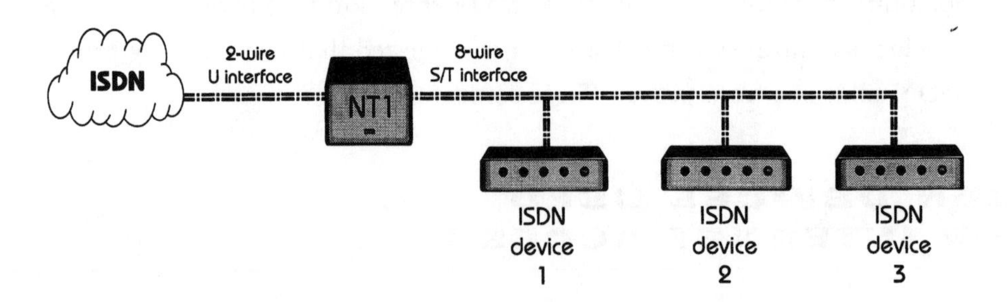

Figure 4.1 NT-1 connecting multiple devices to the ISDN line.

READER'S NOTES

Some devices claim to contain an integrated NT-1. Many of these devices do not support the conversion from U to S/T and, in fact, only have the U interface port (no S/T port). If you buy one of these devices, you will not be able to support additional ISDN devices on your ISDN line such as an ISDN telephone or an ISDN-based videoconferencing system.

TERMINAL ADAPTERS

The terminal adapter (TA) is a device that interfaces a non-ISDN device (for example, a computer) to the ISDN network (Figure 4.2). TAs come in a variety of shapes and sizes. Some are external stand-alone boxes that look like analog modems. Others are cards that fit into PCs or Macintoshes.

External TAs plug directly into an RS-232/communications port on the back of a PC, Macintosh, or Unix workstation. Internal TAs plug into an available slot in the computer. Both types of TAs communicate with the computer via the same Hayes AT command set (for example, ATDT5551212) used by analog modems. This standardization makes them backward compatible with most modem-based communications packages such as ProComm, Microsoft Terminal, or Versaterm.

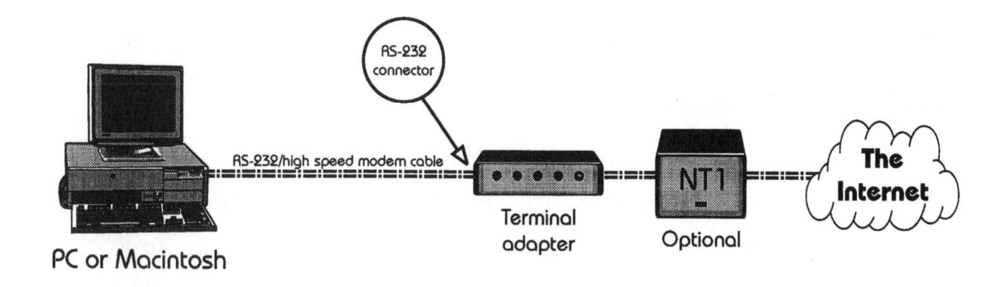

Figure 4.2 Computer connection to terminal adapter.

Some TAs have additional features for Internet users such as built-in analog ports for analog modems, telephones, or fax machines. You can use one ISDN channel to dial the Internet at 57.6 Kbps while using the other channel to talk on the phone or send and receive faxes. You can even buy a TA with a built-in analog modem; these devices can interoperate with both analog modems and digital ISDN devices.

TAs connect to the computer at current maximum speeds of 115.2 Kbps. In the future, speeds up to 230.4 Kbps or 460.8 Kbps will be supported when compression becomes standard on TAs; TAs today generally do not support compression. Prices range from $300 to $1000 depending on features.

READER'S NOTES

When buying your TA, make sure that it supports the V.120 protocol, async-sync PPP/ML-PPP conversion, or both. These are the most common protocols for Internet access, and they will give you the most flexibility in the future to access other types of ISDN-based information services. You may also want to considered getting a TA with an analog port so you can use your old analog modem to connect to non-ISDN information services. Finally, if the product claims that it supports a 115.2 Kbps interface, make sure that it is able to combine the bandwidth of both 64 Kbps channels via either ML-PPP or some proprietary B-channel aggregation protocol. If it does not support channel aggregation, you will only be connecting to the Internet at essentially 64 Kbps, not 115.2 Kbps.

ISDN LAN CARDS

ISDN LAN cards are the second type of device used by ISDN Internet users. ISDN LAN cards interface to the computer very much like network interface cards, such as Ethernet cards, hence their name. Communication between the computer and the card is accomplished via standardized application programming interfaces (APIs) such as NOIS, ODI, or WinISDN, which will be discussed later.

Generally speaking, internal cards can be a little more difficult to work with. The primary problems that arise with ISDN computer cards are lack of available slots, lack of available Interrupt Requests or IRQs, and reliance on the computer for power. The first two problems occur with any new card that needs to be added to a computer. The power issue, however, is a new one that ISDN adds to the equation, and it deserves a short explanation.

Some ISDN cards have external analog ports for plugging in phones, modems, or fax machines. Because the computer powers the ISDN card, and the ISDN card powers the analog interface, the computer must be on to make or receive calls. If you want to place and receive calls even when your computer is off, you may want to buy an external TA for your ISDN device needs.

There are many benefits to using an internal ISDN LAN card. These cards usually support full connectivity over both 64 Kbps channels at the same time. Some people refer to this as channel aggregation or bonding (don't confuse the word bonding with the protocol by the same name BONDING). Another benefit of ISDN LAN cards is that they usually support compression, which means that ISDN LAN card users should be able to experience better than 128 Kbps throughput when accessing the Internet.

READER'S NOTES

When buying an internal ISDN LAN card, make sure it supports the PPP/ML-PPP protocol. This is the most common protocol supported by ISDN LAN cards for Internet access.

ROUTERS

A router is a device that connects LANs to other LANs. Routers are used to connect a collection of computers in a small or medium office to the Internet (Figure 4.3). If 128 Kbps LAN access is what you need for your office, you

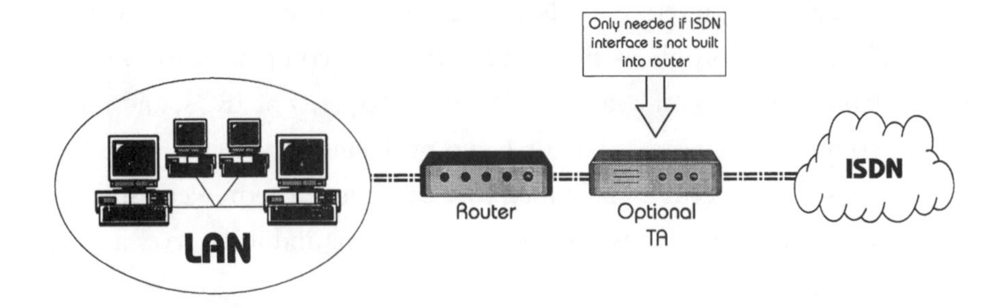

Figure 4.3 Router connecting office LAN to Internet.

will need to buy an ISDN router. Routers can come with a built-in ISDN interface, or they can come with a serial communications interface that can be connected to an external TA. To use a router, you will need a LAN in your office and TCP/IP software running on every computer that will access the Internet. Basic Internet routers cost between $1,500 and $3,000.

ISDN DEVICES USED FOR OTHER APPLICATIONS

Now that we have talked about the types of products used to access the Internet, let's examine a couple of the other types of ISDN products—bridges and video conferencing systems. Although these two types of ISDN products have an important role in ISDN communications, neither is a standard solution for Internet access.

BRIDGES

A bridge is a device that connects either a PC or LAN to another LAN. The problem with bridges is that they are not as selective as ISDN LAN cards, TAs, or routers are about what types of traffic they will send over the link. Bridges may send unwanted packets of information not related to the Internet application across the ISDN line, such as Novell NetWare broadcast packets whereas a router would filter out these packets. This is the main reason why Internet providers do not support bridging.

VIDEO CONFERENCING SYSTEMS

Video conferencing is a very popular ISDN application. You can buy a system for your computer and video conference with people around the world. The problem with video conferencing systems is that they do not support the protocols needed for Internet access, such as V.120 or PPP/ML-PPP.

ISDN-RELATED HARDWARE ISSUES

If you choose to access the Internet via ISDN, you need to make sure your computer has the ability to support data transfer at speeds of 128 Kbps or more. In many cases, your computer may not have the resources needed to have an optimized connection to the Internet, and you will either have to upgrade our computer or some of its components. You may want to read this section so you can determine what upgrades might be necessary to support the speed of ISDN.

COMPUTER REQUIREMENTS

If you are planning to upgrade to ISDN, and you own an older and slower PC (386 or earlier) or Macintosh, you may want to consider upgrading your computer. To truly realize all the benefits of ISDN speeds, you will want to have a computer with the following attributes.

IBM PC

An IBM or compatible PC should meet these minimum requirements:

- Processor: 486 computer or better
- RAM: 8 Megabytes or more
- Serial Port Chip: 16550UART(A)(F)

MACINTOSH

An Macintosh should meet these minimum requirements:

- Model: PowerMac or AV series Mac
- Processor: 68030 or better

- RAM: 8 Megabytes or more
- Operating System: System 7.0 or better

HIGH-SPEED SERIAL PORTS AND EXTERNAL TAS

Many older PCs (386 or early 486 computers) and Macintoshes do not have serial/communications ports fast enough to support the speeds of external ISDN terminal adapters. For this reason, you may have to upgrade your computer's serial ports to realize the speed benefits of ISDN.

For PC users, you can determine what type of serial ports your computer has by executing the program MSD in DOS (do not execute this program from a DOS shell in Microsoft Windows). If you choose the COM Ports menu, a summary of the serial ports in the computer will come up on the screen. The important item you are looking for is the UART Chip Used entry. If the value of this entry is 16550(A)(F) or 16650, you have one of the new serial ports; you should be just fine. If the result comes up with 8250 or something other than 16550, you have an older serial port; you will have to buy a new card for your computer. These cards cost about $50 and can be purchased at local computer stores. They usually come with two high-speed serial ports (16550A UARTs) and one parallel port. Because you already have serial ports in your computer, you will want to disable the existing ports and use the new card for all your computer port needs. The process of installing one of these cards will be discussed in the next chapter.

For Macintosh users, the only way to tell if your Mac has a high-speed serial port is by the computer model. Only PowerMacs and AV series Macs have high-speed serial ports that are capable of communicating at speeds in excess of 57.6 Kbps (that is, 115.2 Kbps or 230.4 Kbps). If you do not own one of these models, the only way to support the higher speed connections is to install a high-speed serial Nubus card in your Macintosh if you have an available slot (Powerbook users will be out of luck here). You will have to pay about $200 for a board with two high-speed serial ports.

HIGH-SPEED MODEM CABLE AND MACINTOSHES

One important requirement of the ISDN Internet access hardware and software is that they support what is called hardware flow control. Hardware flow control allows either of the communicating entities (that is, an external TA and the computer) to stop the other entity from sending any more data over the connection. This control is implemented on two wires (called RTS and CTS) of the RS-232 cable between the computer and the external ISDN TA. Standard Macintosh modem cables do not have the proper hardware flow control wires inside the cable. You therefore have to use a high-speed modem cable for your Macintosh. When upgrading to ISDN, you will want to buy one of these cables if you do not already have one.

GENERAL SOFTWARE NEEDS

You don't have to purchase any special ISDN software for Internet access per se. Most software that you will need will come with the ISDN product you buy. You will however have to obtain a TCP/IP stack to use with your ISDN device. A short discussion of the implications of ISDN on the TCP/IP products you will purchase, as well as the software you will use on the PC to interface the TCP/IP products with the ISDN hardware, will be valuable.

TCP/IP PACKAGES

Many popular TCP/IP packages work with either modems or ISDN devices. These products range from $100 to $500 depending on the types of features they include. Some of the most popular ones are listed in Appendix F. If you are interested in saving some money, you can always use shareware. Many people successfully use products such as Trumpet Winsock or MacPPP, both shareware, to connect to the Internet via ISDN. Refer to Appendix F for information on where to download these software packages from the Internet.

One important issue to address when picking your TCP/IP software is what type of computer interfaces it supports (that is, serial/communications port, Ethernet card, etc.). For ISDN TAs (internal or external), you will need a package that supports PPP and SLIP (PPP being the preferred protocol) to ensure that the software package talks to the ISDN device via a COM port. You will also want to be sure that the software supports the selection of COM port speeds of at least 115.2 Kbps. Some of the less expensive packages only support 57.6 Kbps or less. If you are a Macintosh and MacPPP user, you will want to upgrade to MacPPP version 2.1.2SD or later. This is the first version of MacPPP that will support communication port speeds of 115.2 Kbps and 230.4 Kbps.

For ISDN LAN cards, you will require a package that supports NDIS, ODI, packet, or WinISDN drivers. These drivers are software programs that translate requests between the TCP/IP software and the ISDN card. NDIS is used typically for connecting Microsoft Windows for Workgroups computers to ISDN LAN cards. ODI is used to connect Novell NetWare computers to ISDN LAN cards. Packet drivers and WinISDN drivers have little dependency on network operating systems. If you do not have any networking software packages on your computer (for example, Microsoft Windows for Workgroups or Novell NetWare), you will most likely want to use a packet driver or WinISDN driver if both your ISDN card and TCP/IP package support them.

READER'S NOTES

The cost of the TCP/IP package will increase for ODI or NDIS implementations. This issue should be used as part of the decision process of selecting an ISDN LAN card or terminal adapter (internal or external). It could make a difference of several hundred dollars.

Table 4.1 Types of TCP/IP software drivers for ISDN devices.

ISDN Device Type	Software Drivers Needed
TA	PPP/SLIP
ISDN LAN Card	NDIS/ODI/Packet Driver/WinISDN

WINISDN

WinISDN is a software specification for interfacing Microsoft Windows software products with ISDN cards. This specification has not been widely embraced by the ISDN or TCP/IP software industry; however, a few products do support the specification. A WinISDN driver, to a certain extent, does ease the installation process of an ISDN board. Consult your hardware and software product vendors to determine whether they support the WinISDN standard.

HIGH-SPEED SERIAL PORT SOFTWARE DRIVERS

One of the largest problems facing users of Microsoft Windows 3.x and ISDN is that Microsoft Windows' default communications port driver does not adequately support speeds much higher than 19.2 Kbps. When connecting an external TA to an RS-232/COM port, it is usually wise to upgrade the communications driver that comes with Microsoft Windows 3.x to a faster one. Appendix F contains a list of companies that sell drivers. This software usually costs less than $30.

For users of Windows 95, Windows NT, OS/2, Unix, or Macintosh (PowerMac or AV series Mac) platforms, this driver upgrade should not be necessary because the default serial port drivers can already handle speeds of up to 115.2 Kbps.

SUMMARY

When making your final decision on hardware and software, you need to address the following questions.

HARDWARE

- Do you want an external or internal unit?
- Do you want an analog port for a phone, modem, or fax machine?
- Do you want an integrated modem (for seamless interoperability with services that don't support ISDN yet)?
- Do you want compression?
- What ISDN protocols do you need for your Internet provider (V.120, PPP, ML-PPP, etc.)?

SOFTWARE

- Does the software support 115.2 Kbps and PPP for connection to a TA?
- Does the software support LAN card drivers for connection to an ISDN LAN card?

These questions should limit the options to just a couple of products. At this point, you will be in a better position to choose the final product.

CHAPTER

5

INSTALLING ISDN SERVICE, HARDWARE, AND SOFTWARE

This chapter describes the installation process after the ISDN line and the Internet service have been ordered. Some parts of the installation you may choose to execute yourself; some will be performed by others. As with ordering the service, it is best to approach this subject by breaking down the principal elements as follows:

1. Installing the ISDN telephone line
2. Wiring the premise
3. Installing the NT-1(if necessary)
4. Installing the ISDN hardware
5. Installing the Internet software
6. Bringing up the service

INSTALLING THE ISDN TELEPHONE LINE

After ordering the service from your local telephone company, you should expect a visit from a telephone technician to your office or residence. The purpose of this visit is to complete the ISDN line installation. This entails confirming that the line is operating properly and performing any internal wiring needed, such as installing a new jack for the ISDN line near a computer.

The initial line installation involves locating the ISDN line at the point where the telephone company terminates all its telephone lines to the building. This location is frequently referred to as the "DEMARC" or demarcation point. In a residence, the demarcation point will usually be in the basement or garage. In an office, it will usually be in a closet somewhere in the building, perhaps in the basement. In a business installation, the telephone technician will have to connect the ISDN line from the telephone demarcation point to the telephone closet in the actual office. Once the ISDN line has been set up, the technician will mark the circuit numbers, telephone numbers, or both on the jack where the ISDN line is terminated.

An ISDN line is generally checked by using a device called a butt set. This device is a small hand-held computer that helps the technician verify that an ISDN line is working. The technician will use the butt set to verify that the ISDN line can be initialized properly and that phone calls can be placed and received for data and voice.

To ensure that the data capability of the ISDN line is working, the technician will most likely run what is referred to as a bit error rate test, or BERT, over each 64 Kbps channel of your ISDN line. The BERT test will transmit data out over the line, through the ISDN telephone network, and back to the device to be sure that all the telephone company ISDN equipment is working properly. The technician will verify the voice capability of the ISDN line (if it is subscribed to) by placing a normal phone call to another analog phone.

After the technician is finished testing the line, he or she will be available to perform any in-house wiring that may be required. If you are inexperienced at performing internal wiring, it may be a good idea to use the technician. Tell the technician where your computer will be located, and have him or her terminate the ISDN line at a standard RJ-11 telephone jack as close to the computer as possible. You should request that the line be rechecked with the butt set once the jack is installed at the final destination. This ensures that when you plug in your ISDN Internet device, you are confident that the wiring was performed properly so you can concentrate on installing the ISDN equipment.

READER'S NOTES

Although the telephone technician may indicate that the line is "working" when he or she leaves, this does not necessarily mean that the line is configured the way you need. There is still a very large possibility that the ISDN line parameters are configured in a way which is incompatible with your particular ISDN device or the way you want to use your ISDN device. For example, your device is set up to make two data calls and the switch is configured to allow only one data call. These types of problems can only be found when the actual equipment is installed.

You should expect an installation to take one to four hours to complete. In some cases, the technician may have to leave and return if there is a problem with the telephone network that cannot be solved in short order. Before the technician departs, be sure to ask for the following information regarding the configuration of your service:

- Circuit identification code
- SPIDs (Service Profile Identifiers)
- Telephone numbers

Figure 5.1 serves as a reminder to record the necessary nomenclature and numbers from the telephone service.

Type of Service	________
1B+D (voice and/or data)	________
2B+D (voice and/or data)	________
Switch type (for example, 5ESS/NI-1)	________
Circuit ID	________
SPID 1	________
SPID 2 (if applicable)	________
Telephone Number #1	________
Telephone Number #2 (if applicable)	________
Centrex dialing code (if applicable)	________

Figure 5.1 Line configuration information form.

WIRING THE PREMISE

Even if the telephone technician does the internal wiring, it may be useful to go over a few basics about ISDN wiring here. This information will also be helpful if you plan to re-do your wiring in the future.

ISDN runs over the same standard twisted-pair copper wire that analog phones use. In almost all cases, the standard copper telephone wiring in your building does not need to be changed, unless it is very old and deteriorating. If you are adding an ISDN line as a second or third line and you live or work in a newer home or office, you may not have to install new cable because these buildings are usually prewired to accommodate multiple telephone cables (up to three in some cases) from a single jack. If no spare pairs exist, you will have to run a dedicated wire pair to the room or office where the ISDN device will reside.

You will want to terminate the ISDN line at an RJ-11 jack(standard for analog phones) in the room where the ISDN Internet device (terminal adapter or ISDN LAN card) will reside. If you are installing new cable, you may have to install a new jack specifically for the ISDN line. Wire polarity is not an issue with the 2-wire U interface, so the two wires can go in any order in the jack. The only requirement is that the two wires occupy the center two pins on the RJ-11 jack.

At this point you are ready to begin plugging in the actual hardware. Figure 5.2 illustrates the different cable interfaces that you should be familiar with.

You can connect either an NT-1 or an ISDN Internet device that has an integrated NT-1 to the telephone line with a standard 2-wire or 4-wire cable. If you are using an ISDN Internet device that has an integrated NT-1, your wiring task is complete and you can move on to the next section of the chapter.

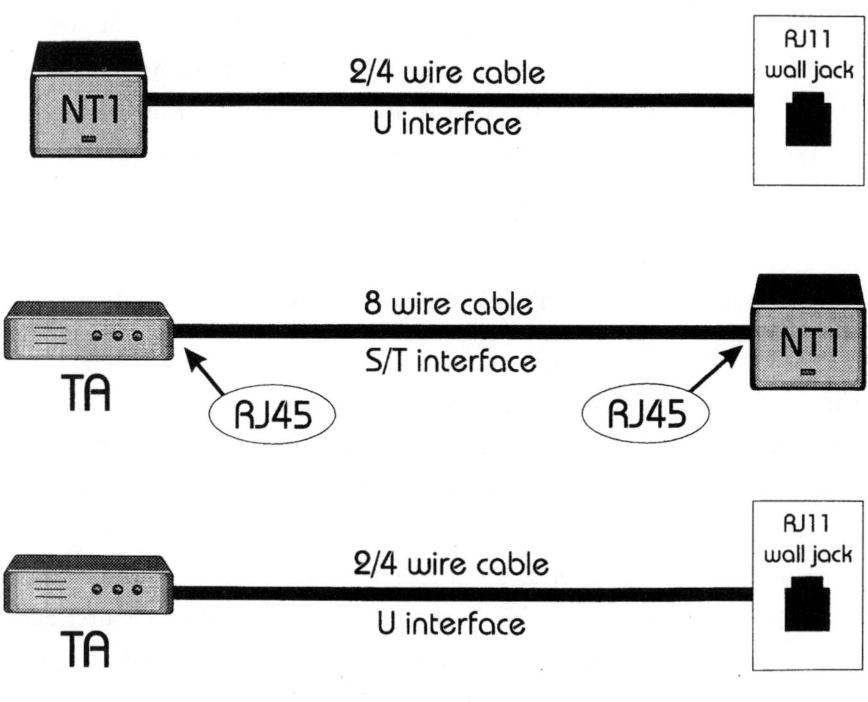

Figure 5.2 NT-1 and ISDN device in same room.

READER'S NOTES

Do not plug an ISDN device that supports the 8-wire S/T interface into the 2-wire U interface. You will need to plug this type of device into an NT-1, and the NT-1 will plug in turn, into the 2-wire U interface.

If you are using an NT-1, you will also need to do the wiring between the NT-1 and your ISDN Internet device. This will involve using 8-wire S/T interface cable instead of the 2-wire U interface cable. If the NT-1 and the ISDN device are in the same room, the wiring is simple. You connect the NT-1 S/T interface port to the S/T jack on the ISDN device with 8-wire cable (usually provided with the NT-1 or the ISDN Internet device). If the two devices are in different rooms, you will need to pull 8-wire cable from the room where the NT-1 resides to the room where the ISDN device resides. In this situation, you will be using RJ-45 jacks (for 8-wire plugs) instead of RJ-11 jacks. Polarity on the S/T interface is very important, so be careful to keep the wires in the same order as they start at the NT-1. Needless to say, it is much easier to have all your ISDN equipment in the same room if at all possible.

The following Figure 5.3 will help explain the basic wiring configurations described above.

The final topic is the role of the termination resistor. When using the S/T interface, you are supposed to properly terminate the S/T wire with a 100 ohm resistor. Figure 5.4 illustrates this point. In many cases, the NT-1 will have a built-in termination resistor to take care of this for you. If you are using only one or two devices and these devices are all in the same room, you will not have to concern yourself with terminating the S/T interface. If you want to wire up multiple rooms with ISDN, you may need to consider putting a termination resistor at the end of the line.

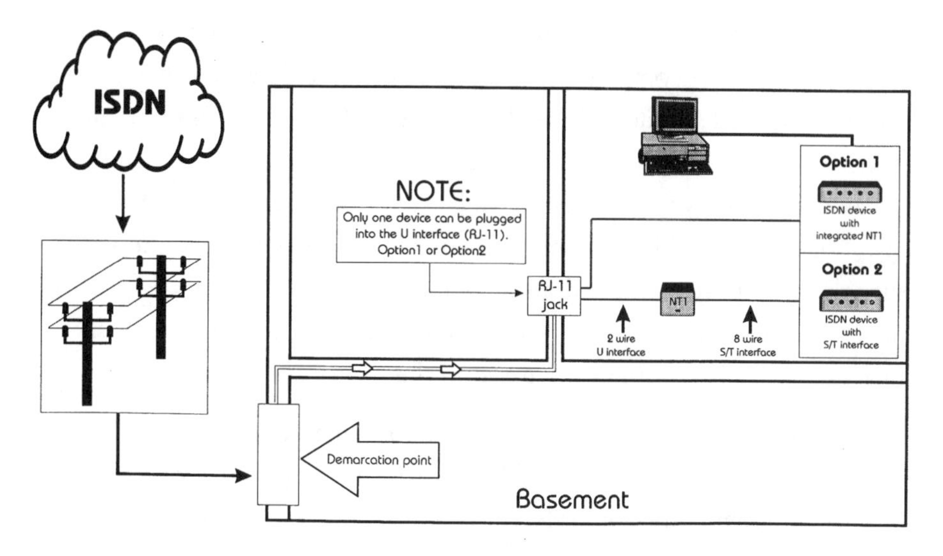

Figure 5.3 NT-1 and ISDN device in different rooms.

INSTALLING THE NT-1

If you will be using multiple ISDN devices on your ISDN line, an NT-1 will have to be installed. The installation of the NT-1 will give you your first opportunity to discover any wiring mistakes by way of its diagnostic capability.

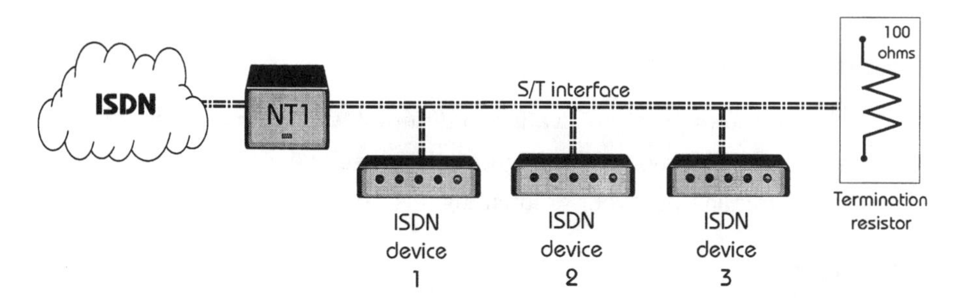

Figure 5.4 Termination resistor terminating the S/T interface.

An NT-1 will have several telephone jacks. One of these jacks will interface with the telephone company over the 2-wire U interface. Another one or more of these telephone jacks will interface with the ISDN Internet devices over the 8-wire S/T interface. In the case of NT-1 Plus devices (defined earlier in chapter 4), you may also have jacks for other devices, such as analog telephones. You will have to refer to your NT-1 documentation to make sure that you have connected all the wires to the correct jacks. Note that the S/T interface port cannot be connected and verified until the ISDN Internet device is set up, as discussed in the next installation section.

An NT-1s usually has at least three indicators that verify it is functioning properly: power, U interface, and S/T interface. These indicators may have slightly different names on different products; however, their intent is the same. The power indicator will verify that the unit is receiving power. The U interface indicator will verify that the interface to the telephone company is activated. The S/T interface indicator will verify that the interface from the NT-1 to the ISDN Internet device is activated. You can verify that the NT-1 unit is installed properly by checking that all the indicators show activation (that is, power is on, U interface is activated, and S/T interface is activated). It can take up to 30 seconds for the U interface to activate after being plugged in. If it does not activate in that time frame you can begin to look for a problem.

READER'S NOTES

Each NT-1 shows proper activation in different ways. Some units turn green lights on, some turn red lights off. Consult your NT-1 product literature to determine how your product signals activation.

If all the indicators are activate, you can assume the NT-1 is successfully installed and you are ready to install the ISDN Internet device. If any of the indicators are not activated, you will need to verify your wiring and make sure all the telephone cables are firmly connected to the NT-1. If this does not help, you will have to call the telephone company to verify that your service is, in fact, activated.

INSTALLING YOUR ISDN INTERNET DEVICE AND ASSOCIATED SOFTWARE

This section discusses the general process of installing the ISDN Internet device and any software that comes with the device. You will need to refer to the product manual for any detailed installation information.

PREPARING TO INSTALL YOUR ISDN INTERNET DEVICE

Before you start connecting TAs or plugging in computer cards, there are two things you may need to do in preparation. First, you will need to install a new serial port card if your computer does not have a high-speed serial port (16550 UART) and you plan to use an external TA. Second, you may want to install a new communications port driver if your computer is running Microsoft Windows 3.X.

INSTALLING A NEW HIGH-SPEED SERIAL CARD IN YOUR PC

To install your new high-speed serial card, you will need to power off your computer, open it up, and determine whether your existing serial ports (that is, COM1, COM2) and parallel port (that is, LPT1) are part of an existing card or part of the motherboard. If they are part of an existing card, such as a multipurpose disk controller and I/O card, you will need to remove the card so you can examine the jumper settings on the board.

READER'S NOTES

You should not disconnect the cables to the disk drives or the RS-232 ports unless you are experienced at setting up these types of cards or you have documented the exact cable set-up.

Once the card is removed, you will want to set the appropriate jumpers on the card to disable COM1, COM2, and the parallel port LPT1. The manual for the multipurpose card will explain which jumpers to change. Once you have disabled the three communications ports, you can put the card back in the computer. If an RS-232 connector bracket is being used for the second COM port, you may want to consider removing it because it is wasting an available slot.

If the communications ports are part of the motherboard, they can most likely be disabled by commands in the BIOS SETUP program. The BIOS SETUP program can usually be engaged by pressing either the DELETE or the F1 key immediately after the computer boots up. If neither of these keys works, you will need to consult your computer documentation for information on how to access the BIOS configuration.

READER'S NOTES

Be very careful when making modifications to your system BIOS. One erroneous change can disable your computer.

There should be a place under one of the menus that allows you to disable COM1, COM2, and LPT1. Once you find the option and set it, you should follow the instructions to save the BIOS settings to CMOS. The computer will reboot after you save the settings. When the computer comes up, you might want to run MSD in DOS to verify that there are no longer any active COM ports.

At this point, you are ready to install the new card. Although this card will have jumpers on it, they should be set to the default values for a normal computer. Simply locate an open slot and slide the card into the computer chassis. If you want to connect your ISDN Internet device to the second COM port (because the mouse uses COM1), you will need a second open slot right next to the high-speed serial card to plug in the connector for the second RS-232 port.

After completing your high-speed serial card installation, you may want to install the communications port driver upgrade software for Microsoft Windows 3.X. The software will come on one disk and should install in a matter of seconds.

READER'S NOTES

This software will replace the existing comm.drv file that comes with Microsoft Windows. If you are using Remote Access Service (RAS), or other software that relies on the original "comm.drv" driver, you may not be able to install a new communications driver. In this situation you will need to upgrade to Windows 95, Windows NT, or OS/2 warp to get the full performance of ISDN.

INSTALLING AN EXTERNAL TERMINAL ADAPTER

To install your TA, you will need to connect an RS-232 cable from the communications port on the back of the computer to the RS-232 port on the TA, connect the ISDN line into the back of the TA, and connect the power supply to a wall outlet. Remember to use a high-speed modem cable if you are a Macintosh user. If the TA supports the S/T interface, the ISDN line will be an 8-wire RJ-45 cable from the S/T port on an NT-1. If the TA supports an integrated NT-1 or U interface (these are one and the same), the ISDN line will be a 2-wire RJ-11 cable from the telephone jack on the wall. Figure 5.5 shows these two configurations.

After plugging in the TA, your next step is to configure the software on the unit. This can be accomplished in four different ways, depending on the unit. The information could be entered by Hayes AT commands (for

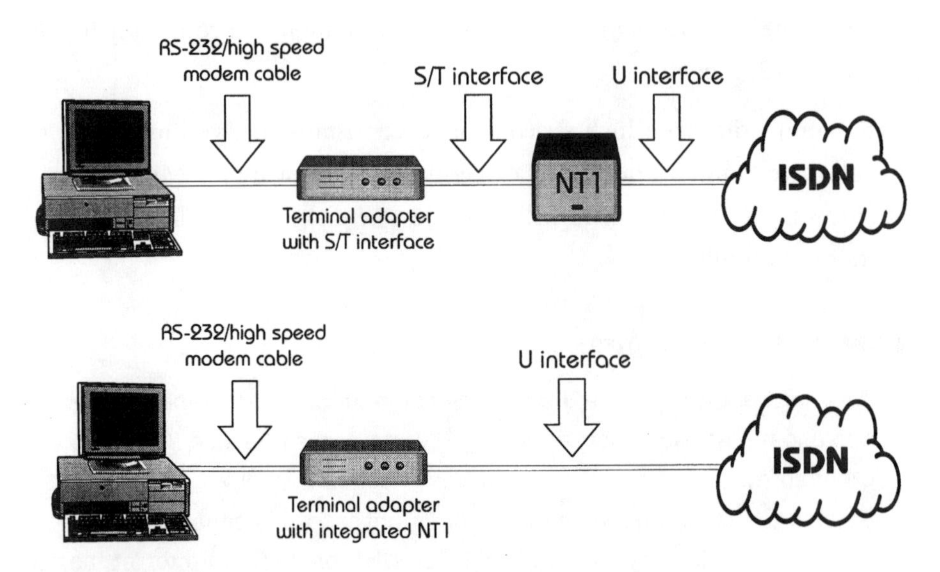

Figure 5.5 Connecting external terminal adapters.

example, ats51=012075551212); it could be entered via a Microsoft Windows- based program that runs on the computer; it could be entered via a front panel keypad; or it could be entered via a menu program that runs under a terminal emulation program such as ProComm or Microsoft Terminal. You will have to consult the product literature to determine which method applies to your product.

Once you determine the method of entering the information, your next step is to enter the ISDN information provided by the telephone company—namely the ISDN switch type, the SPIDs, and the telephone numbers. The switch type in most cases will need to be set to National ISDN-1 (NI-1). In the United States, National ISDN-1 is by far the most widely standardized switch configuration. Other possibilities for switch type are AT&T Custom and DMS-100. You will need to enter the telephone number(s) and SPID(s) as well. The telephone number will be a standard 7- or 10-digit number. The SPID(s) will be a 9-20 digit identifier that typically comes in the following format: <prefix><directory number (DN)><suffix>. The directory

Table 5.1 Sample SPID Formats.

Switch Type	SPID Format
AT&T 5ESS	01-5551212-000
DMS-100	901-5551212100
SSC EWSD	314-5551212-0100

number and telephone number are one and the same. Table 5.1 shows sample SPID formats for three different ISDN switch makers. Depending on how the telephone service was ordered, there may be one or two SPIDs and DNs. If the switch type is AT&T Custom, there may be no SPIDs.

Other parameters that may need to be entered are as follows:

- Data call type (either 64K or 56K; the latter is needed for long distance phone calls)
- Protocol (V.120, async-to-sync PPP/ML-PPP conversion, etc.)
- Bit rate (57.6 Kbps or 115.2 Kbps)
- Flow control (usually set to hardware flow control)

To further illustrate the process of configuring an ISDN terminal adapter, the following screen captures (Figures 5.6 and 5.7) are shown as you would see them on your computer.

As with most products, these screens are self explanatory and step you through the process of properly configuring your device. It should be noted, however, that if you enter the incorrect information the device will not be able to identify the error.

INSTALLING AN INTERNAL TA OR ISDN LAN CARD

ISDN cards have many of the same installation procedures as other types of computer cards. As always, you will have to open the computer and slide

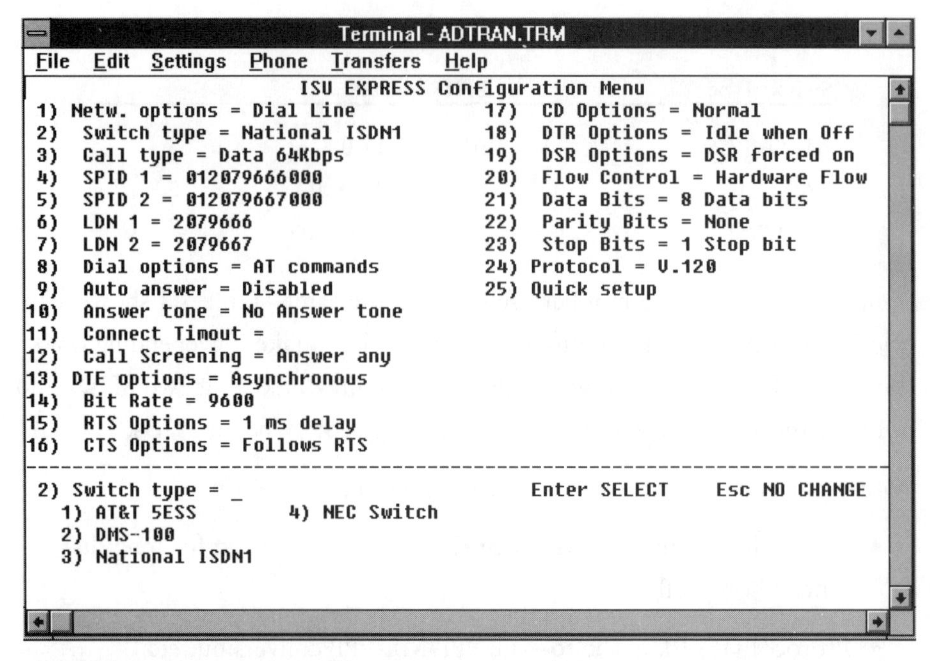

Figure 5.6 Sample configuration screen.

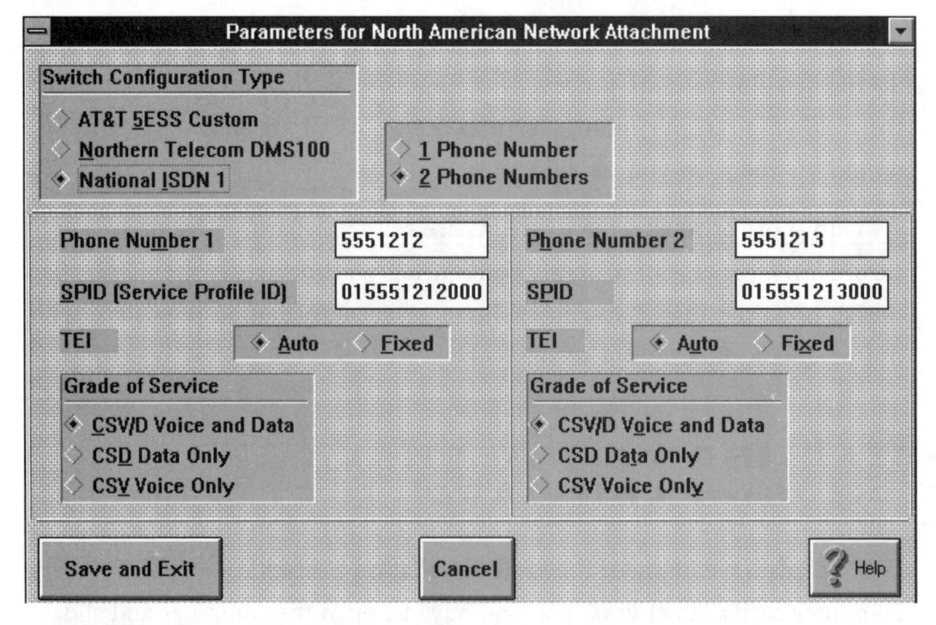

Figure 5.7 Sample configuration screen.

the card into an available slot. If the card has jumpers on it, you will also have to determine the proper jumper settings that will allow the card to function properly. The most common jumper types set the IRQ number, memory address space, and COM port (only for internal TAs). The jumpers come already set to default values that will work well in your computer. If you have a lot of other cards in the computer, such as sound cards and CD-ROM cards, you may have to try a couple different combinations of the jumpers to get your ISDN card working. Internal TA COM port jumpers may have to be moved to avoid conflict with existing COM ports already being used on the computer. You should also be prepared to disable existing COM ports as described earlier to get your internal TA to work. Because of the possible iterative process of setting jumpers, you should leave the computer cover off until you have verified that everything in the computer is working and the ISDN computer card is installed properly.

The next step is to power up the computer and install the software that comes with the card (every card will have software). During this software installation, you will be prompted for the configuration information, such as telephone company parameters (for example, switch type, SPIDs, and telephone numbers). Board parameters (for example, IRQ, memory location, I/O location, etc.), and in the case of ISDN LAN cards, computer interface software (for example, NDIS, ODI, WinISDN, etc.) will also be required. Should you not have all the information you need to complete the installation, you can run the installation program again if necessary.

The details of entering the telephone company parameters are covered in the previous section on installing TAs. Other parameters that may need to be entered are as follows:

- Data call type (either 64 K or 56 K, the latter needed for long distance phone calls)
- Protocol (PPP/ML-PPP for ISDN LAN cards, V.120 or async-to-sync PPP/ML-PPP conversion for internal TAs)

- Bit rate (56, 64, 112, or 128 Kbps for ISDN LAN cards, 57.6 or 115.2 Kbps for internal TAs)

SPECIAL ISSUES FOR ISDN LAN CARDS

ISDN LAN cards generally interface with the computer via special software device drivers. These software drivers allow applications on the computer, such as TCP/IP stacks, to communicate with the ISDN card. The primary four interfaces were introduced in the last chapter: NDIS, ODI, packet driver, and WinISDN. When installing your ISDN card, you will need to select one of these drivers to use. Not every card supports every type of driver, so you will have to choose from what is available. You will need to pick one that is supported by your TCP/IP software as well.

VERIFICATION OF ISDN INTERNET DEVICE INSTALLATION

Once all the parameters are entered and saved in memory, the next step in the installation is to verify that the ISDN device is installed properly. This can be accomplished in a variety of ways. The following lists a couple of the steps of this verification:

- Verify that the ISDN device has initialized itself with the telephone network
- Verify that you can place and receive data calls, voice calls, or both

VERIFICATION OF ISDN INITIALIZATION

Most ISDN devices will attempt to initialize themselves with the ISDN telephone network on power up. By verifying the success or failure of this initialization, you can begin to determine if the device and the ISDN line are working properly. How to verify this initialization is product dependent; it should be documented in the product manual. Most ISDN devices provide one of the following indications: a message on a display, an LED on the front panel, or a log entry in a computer file.

Typical ISDN devices initialize immediately, but some may require up to 30 seconds to initialize; be patient if it does not happen immediately. If the ISDN device is a computer card, the device may not initialize until the computer software is loaded into memory. If the ISDN device does not seem to initialize, there are a few things you can try to correct the situation:

1. Check the connection of the ISDN Internet device to either the RJ-11 wall jack for devices with integrated NT-1s or the S/T interface jack on the NT-1 for devices that support the S/T interface.
2. Power cycle and/or remove and reconnect the telephone cable to the unit.
3. Verify that the SPIDs are stored in memory and that they are correct.

If you have tried the three suggested steps and the unit still does not initialize, you can conclude that there is a problem with either the ISDN line or the ISDN device.

READER'S NOTES

Some devices have more effective techniques than others for providing a clear indication of ISDN line initialization. If your device is one that does not provide a definite indication of initialization, you will have to resort simply to placing a test call. Uninitialized devices cannot place outgoing calls.

VERIFICATION OF PLACING AND ACCEPTING DATA CALLS

The next operation to perform is a test call to your ISDN Internet provider. This test call will verify two things for you. First, it will further confirm that your ISDN line is working properly (verification of initialization was only the first part). It may also verify that the settings on your ISDN device are compatible with the Internet provider equipment (for example, V.120, PPP, 64 Kbps, 128 Kbps, compression, etc.). Some of these settings can only be verified when the actual TCP/IP software is running. If you can successfully place a call to your Internet provider, you are well on your way to setting up your ISDN Internet service.

For TAs (external and internal), placing this test call is relatively easy—the procedure is the same for modems. To accomplish this you will be using your communications software such as ProComm, Microsoft Terminal, or Versaterm. Once you have configured the software for the proper COM port, port speed, and hardware handshaking, you will type ATDT followed by the number you want to dial. The response from the TA will be "NO CARRIER," "BUSY," or "CONNECT" depending on the outcome of the call. If you get a message that says something like "CONNECT 57600" or "CONNECT 115200," you know the outgoing call was successful. If you get the "NO CARRIER" message you probably dialed a wrong number or there might be something wrong with your configuration. Some TAs also use this response when the dialed number is busy. If you get the "BUSY" message the dialed number is busy.

READER'S NOTES

If you are currently a modem user, you probably use the telephone tones you hear when dialing to diagnose your connection to the Internet. If you don't hear dial tone, you know that your modem is not connected properly. If you hear a busy signal, you know your Internet provider has no available modems to call into. This same technique cannot be used with ISDN devices because there are no sounds on an ISDN data call. You will have to use diagnostics from your ISDN devices to determine the problem.

To place a call from an ISDN LAN card, you will generally need to use a special software program that comes with the card, which manually places or terminates a call. This program will be documented in the user manual for the product. Success or failure of a call via an ISDN LAN card will be verified by either a log file stored on the computer or a message returned from the special call-placing program.

ISDN CAUSE CODES

To help you debug failed ISDN call attempts, some devices provide you with call status information that the ISDN network returns to the ISDN device. One of the pieces of information received is called a cause code, which is similar to a diagnostic response. A cause code is sent by the ISDN network to explain why something happened the way it did. Typically, cause codes are associated with terminating an ISDN call. For instance, if you place a call and the line is busy, you might get a cause 17, which means "user busy." Other examples of cause codes you might receive are 16, 34, 44, or 18, which stand for normal clearing, circuit channel congestion, requested channel not available, and no user responding.

By looking at the values returned from the ISDN network, you can begin to understand why a call is not going through. Table 5.2 lists the cause codes you might receive and how they relate to the configuration of your ISDN device.

Table 5.2 Typical cause codes.

Code	Meaning	Explanation
16	Normal clearing	This means that you failed in the B-channel protocol negotiation either by sending the wrong protocol or sending the wrong name and password for authentication. For example, if you configured your unit for V.120 and the box you were calling was configured for PPP you might get this cause code.

17	User busy	This is the same as getting a busy signal on a modem.
18	No user responding	The ISDN device you are calling is not answering.
34	Circuit/Channel Congestion	You are trying to place a call when you already have two channels connected.
44	Requested channel not available	This cause is similar to cause 34.
100	Invalid information element contents	You might get this cause if you entered the wrong SPID value in your unit.

VERIFICATION OF VOICE CAPABILITY

If you have selected an ISDN device that has an analog port for a telephone or fax, you may also want to try a voice test call. If voice calls can be placed in both directions, you will confirm that the ISDN line is configured properly for voice operation. If the voice line has other features such as additional call offering, conference, call forwarding (voice mail), or call transfer, these features should be tested as well. The ISDN device documentation should guide you through these procedures.

INSTALLING THE INTERNET SOFTWARE

Once the ISDN portion of the installation has been completed (wiring and installation of device), you can then proceed to configure any other software or external interfaces that might be necessary for operation of the ISDN device (that is, network connection, communications software, TCP/IP software).

INTERFACING TCP/IP SOFTWARE WITH ISDN TAS

If you are using an ISDN TA, the TA will look like a modem to the computer. Any software packages that work with modems therefore should work with ISDN TAs. In the case of TCP/IP software, it must support PPP and SLIP protocols. The details of installing TCP/IP software are not discussed in this book. Please refer to Paul Gilster, "The SLIP/PPP Connection" published by John Wiley & Sons, for additional information on how to set up your Internet software. There are two points worth mentioning about your software installation, however. First, you want to make sure that the port speed of the software is set to the same value as the ISDN port speed (usually 115.2 Kbps). Second, you want to enable hardware flow control. ISDN just seems to work better when hardware flow control is enabled.

INTERFACING TCP/IP SOFTWARE WITH ISDN LAN CARDS

If you are using an ISDN LAN card, you will probably be installing TCP/IP software that works on LANs. You will have to configure the software driver interface in the TCP/IP software to get it to work with your ISDN card. The value of this parameter will have to be NDIS, ODI, packet driver, or WinISDN, depending on how you configured your ISDN LAN card.

MAKING THE FIRST CONNECTION AND TROUBLE SHOOTING

Once all the TCP/IP software programs and associated drivers are installed (that is, IP information entered, telephone numbers, script files edited, etc.), you should be ready to connect to the Internet. Engage the software to make a connection, and after a couple of seconds you will be on the Internet if everything is working properly and no mistakes were made when configuring the ISDN device or the software. Let us assume that there still may be issues with your connection. At that point the trouble shooting process begins. To assist you through this process we have provided troubleshooting hints.

TROUBLESHOOTING HINTS

To assist in your troubleshooting, this summary of issues may be helpful. The list is a sampling of some of the more common problems that may occur when one installs an ISDN line.

1. Your ISDN device does not initialize with the ISDN switch
   a. you entered the SPIDs incorrectly
   b. telephone company gave you wrong SPIDs
   c. telephone company programmed switch incorrectly
2. You can place voice calls but no data calls or vice versa
   a. you did not order voice and/or data service on your line
   b. telephone company programmed switch incorrectly
3. You can only place one data call instead of two simultaneous data calls
   a. you did not order an ISDN line configuration that supports two simultaneous data calls
   b. you programmed your ISDN device to place only one data call
   c. telephone company programmed switch incorrectly
4. Call goes through but gets dropped within 30 seconds
   a. protocols are set to different values on each side of the connection
   b. user name or password entered incorrectly in TCP/IP software
   c. data rate on your ISDN device might need to be set to 56 Kbps instead of 64 Kbps
5. Call does not go through at all
   a. your ISDN device has not initialized with the ISDN switch
   b. you dialed the wrong number
   c. Internet service provider has no available capacity at the moment
   d. data rate on your ISDN device might need to be set to 56 Kbps instead of 64 Kbps

e. interface between TCP/IP software and ISDN device not configured properly

6. Call goes through, stays connected, but Internet access does not work

a. you entered in wrong IP address, or Domain Name Server (DNS) server address

SUMMARY

The purpose of this chapter was to discuss many of the essential elements in the installation process of both hardware and software to connect to the Internet via ISDN. It is understandable that to some Internet enthusiasts the process discussed in chapters 4 and 5 could appear somewhat daunting. However, if you have gone through the experience of setting up a computer, installing a modem, and installing communications software you should be able to tackle ISDN with comparable success.

It is true that at its current stage of implementation there are many segments of the process to deal with. When and if they appear to exceed your skill level and/or experience you will find lots of help through your Internet provider and ISDN equipment manufacturer. Both Internet service providers and ISDN product manufacturers are keenly aware of the need for further end-to-end simplification of the overall process. Because of the intense interest in utilizing the increased bandwidth of ISDN communications, if only for Web access, you can expect seamless and simplified procedures to be forthcoming. The features of the so-called "Plug and Play" technology, soon to be employed widely in computers, will undoubtedly contribute to the simplification process.

Using the information provided in this book will hopefully guide you to the best performance you have ever experienced on the Internet.

To summarize many of the procedures described in this chapter, Figure 5.8 presents a flow chart of a typical installation. The flow chart describes the process of installing an external stand-alone ISDN terminal adapter to a PC with a high-speed serial port.

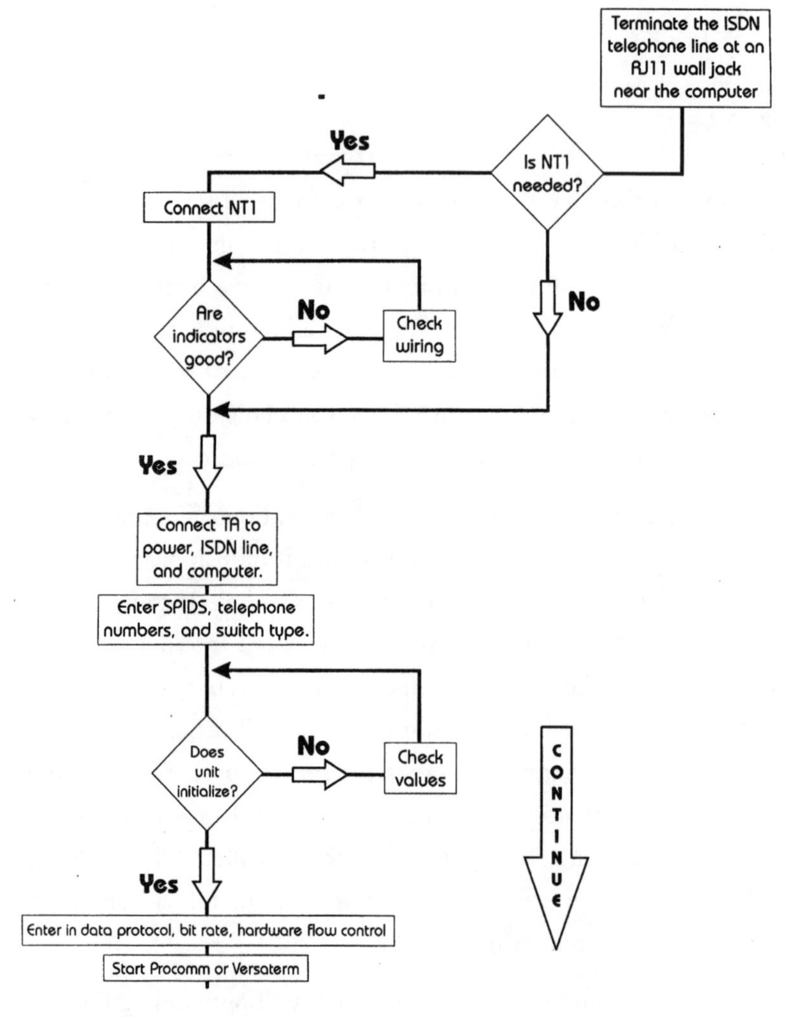

Figure 5.8 Installing an external terminal adapter.

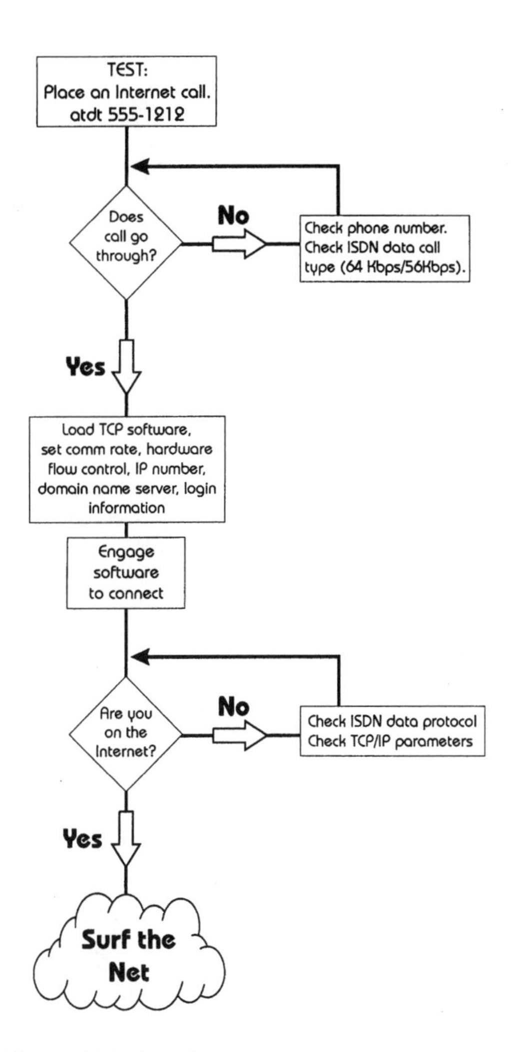

Figure 5.8 Continued

PART

2

ISDN TECHNICAL INFORMATION AND BACKGROUND

CHAPTER

6

DATA COMMUNICATION NETWORKS AND ISDN

The evolution of data communication networks has been a series of improvements in the speed, and access, of data systems. The first step was the development of communication methods. The next was to create fast methods of long distance transmission. The current step is continual improvement in speed and commonality of access.

COMMUNICATION IN THE ELECTRONIC WORLD

What do data communication networks have to do with ISDN and the Internet? It is very important to back off a bit from the use of high-speed access to the Internet and examine what actually is part of communication. This situation parallels what many beginning Internet users encounter with Mosaic. Mosaic relies on "helpers" to decipher various types of multimedia files. If you access a file, it is transferred and then a search for the appropriate helper program is made to decode the data. If the helper program does not exist, the data cannot be used. Thus, no communication has taken place (although the transfer may have taken quite a long time).

THE SEVEN STEPS

Let's apply the seven steps of communication to the electronic world and that of the Internet, specifically. These seven steps are creation, translation, transmission, reception, identification, use, and storage. Creation remains the same for all environments. Translation and transmission are actually a bit easier to talk about in the electronic world because they are better defined.

TRANSLATION FOR DATA COMMUNICATIONS

In the electronic world, translation is the process of encoding (or decoding). There may, or may not, be a preliminary human-oriented translation form.

Translation (and transmission/reception) rely on standards. In the non-electronic world, standards take the form of rules (such as grammatical requirements or spelling orthodoxy) or aesthetic groups (which are often changed with great reluctance by those who have mastered the previous versions). Similar conflict occurs in the electronic world. One user wants to encode in TIFF, another in PICT, a third in their own proprietary format. Identification is needed for the translation part at the reception side of the message.

TRANSMISSION

Transmission occurs when translated data are sent from one point to another.

Let's take a simple case. We have a very good transmission medium, but the line is not in constant use. In this case, it is necessary to mark the beginning and the end of the data. This is called *framing* (similar to a painting being enclosed by a picture frame). The next step is to allow for identification (and, for simple cases, correction) of errors in the transmission. This can be done by various mathematical means. Often, in the digital world, it is done with a frame check sequence (FCS). Framing can also incorporate methods of simple correction.

Several additional factors contribute to the design of a transmission protocol. One is that data are normally split into separate sections. These sections may be called *frames* or *packets*. Once the data have been split, it is now necessary to identify sequencing of the parts of the data. This allows for identification of missing, or erroneous, sections. (It may also allow retransmission of the bad packets.)

Storage

Storage is, in a way, a manner of reception without translation or use. All of the aspects of transmission occur—creation, translation, and transmission. However, storage indicates that, upon reception, the data are placed in a form that can be transmitted again at a later date. An example in the physical world is a printed book, a letter, or even a videotape. Examples in the electronic world include magnetic disk storage.

Reception

Reception is the act of decoding the transmitted data back into "pure" data that can be used by applications (or higher layers of protocols). Minimally, this is a decoding of the physical medium signals into a form that can be translated. In most circumstances, it will also involve interactions with the transmitting side to verify the correctness of the data.

Note that the process of transmission and reception cannot be separated in form. What the transmitting side sends, the receiving side must expect. In addition, some form or method of *routing* of the data will exist. Routing is the path the data follow. In a simple case, such as a direct electrical connection (for example, in an intercom set-up or a direct video link) this path is a method for the physical medium to join the transmitter and receiver. In more complicated cases, there are many different ways and paths between the transmitter and the receiver. This is where signalling methods become involved. Signalling allows different paths to be used between the point of origination and termination of the data transmission. Chapter 7 will look into how signalling and transmission methods have evolved.

Translation at the Receiving Side

At the receiving side, translation takes place at the application layer. (Data may pass through intermediate processes on the way.) As mentioned in the explanation of translation on transmission, the form of the data must be identifiable. Identity can be established via embedded codes, naming conventions, or as part of a binary format. Most Internet tools will use particular applications to deal with different formats. A graphical interchange format (GIF) file will have its own interpreter. A hypertext file will have another interpreter. In many situations, there will be an interactive situation where transmission takes place from the receiving side upon translation of the data.

Usage of Data at the Receiving Side and an Example

Let's examine a potential scenario with use of the Internet. A connection has been made with an Internet access node. The user wants to use Eudora to retrieve mail. He or she starts the application. Many of the connection parameters have already been installed as part of the preferences file. A command to "Check Mail" is issued. A message is sent to the server, which responds with a log-in message. The user ID is sent back to the server, which responds with a request for a password. The user sends the password. After checking, the server notifies the user as to the status of his or her mailbox.

THE BEGINNINGS OF ISDN

The organization chosen to develop an international standard, which would allow integrated services on a digital network, was the International Telegraph and Telephone Consultative Committee. This group, better known as the CCITT, exists as a working group under the International Telecommunication Union (ITU), which is under the supervision of the United Nations. Currently, it is known as the ITU-Telecommunication Standardization Section (ITU-T).

In 1968, a special group within the CCITT was given the task of making sure that standards were designed to employ digital encoding techniques for voice use. However, the group was also instructed to look at all aspects of digital technology. This group, now called Study Group XVIII, prepares recommendations that are to be used as standards by countries in their development of ISDN. Note that these are only recommendations. The recommendations themselves allow for many different options. Some of this flexibility is based on the original idea for ISDN—allowing a smooth transition from existing networks to a global integrated digital network.

Originally, the group met to approve recommendations every four years. (Work on the recommendations continued within task groups in between voting meetings.) Thus, recommendations were approved in 1968, 1972, 1976, 1980, 1984, and 1988. After 1988, it was decided that due to various factors (one of which was that there was an accelerating need for more specific recommendations in shorter intervals), the recommendations would appear as they were ready. During the period 1968–1988, volumes of recommendations were published. Each four-year's recommendations were printed as a set of volumes in a distinct color. 1984, whose recommendations were printed with a red binding, was the first year that a sufficient core of recommendations was produced to allow equipment manufacturers and network providers to create real ISDN products. 1988's blue books detailed the basics sufficiently to get the process going internationally.

SUMMARY

This chapter has been primarily about how ISDN came into being. There has been a continuing demand for faster and more reliable information, which has led to the development of new technologies. The next chapter will introduce ISDN architecture.

CHAPTER

7

THE ARCHITECTURE OF ISDN

As mentioned in the previous chapter, the development of ISDN recommendations is the responsibility of the ITU-T, particularly Study Group XVIII. The first set of ISDN-specific documents was released as part of the red books in 1984, and a more complete (sufficient for implementation) set was released as part of the 1988 blue books.

This does not mean, however, that the world waited patiently for the ITU-T (then CCITT) to bring forth the recommendations. Getting a committee to agree to specifics is always a long process—although it may be a little easier in technical areas than in the political arena. It is even more difficult and time consuming when the group is an international one.

THE ITU-T ISDN I-SERIES RECOMMENDATIONS

The ITU-T's recommendations are mostly released as part of a group called the I-series. These recommendations have the following groupings, as seen in Table 7.1.

Many of the I.400 documents are cross-referenced within other series also. Thus, I.441 is also known as Q.921, and I.451 is known as Q.931. The series are meant to be inclusive of certain categories of recommendations. The I-series is meant for ISDN; the Q-series is primarily for signalling recommendations. Thus, overlapping situations have caused the issuance of the same recommendation in different series. In the early 1990s, the decision was reached to stop the double naming. Thus, ITU-T Recommendation Q.933 (which concerns ISDN signalling for support of switched frame-relay connections) does *not* have a corresponding I.4xx reference number.

Most of the ITU-T recommendations do not have direct use for people wanting to use ISDN for access to the Internet. There are three areas, however, that do provide information that a general user of the network needs. These are the areas of architecture, service provision, and specific protocol capabilities.

Table 7.1 Categories of ITU-T I-Series documentation.

Series	Purpose
I.100	General structure
I.200	Service recommendations
I.300	Overall network aspects and functions
I.400	ISDN user-network interfaces
I.500	Internetwork interfaces
I.600	Maintenance principles

ITU-T ISDN ARCHITECTURE

The architecture is primarily covered in the "general structure" documents, or the I.1xx series. The first documents (I.11x) discuss the general aspects of how ISDN will be described. The next subset (I.12x) gives a general description of ISDN. The remaining documents cover maintenance and modeling aspects and introduce the concept of services.

ITU-T Recommendation I.120, in particular, is titled "Integrated Services Digital Network" and covers much of the basic meaning of ISDN; it discusses what each of the four words that make up the acronym mean. Digital and network, in part, explain themselves because they do not refer to anything special within ISDN relative to other digital or network situations.

Integrated is actually the key word. Voice and non-voice are both to be supported on the same network. Switched and non-switched (dedicated) connections are to be allowed. Switched connections should allow for both circuit switching and packet switching (and other technologies in the future).

Support of integration requires a type of "intelligence" in the network, which allows examination of the use of the network and management of resources. Integration is an area on which it has proven to be the most difficult to gather consensus. Integration specifically allows for, and supports, interworking capabilities. It also provides identification information needed for user equipment to know the precise protocol being used. Because of the lack of consensus, the task of the user in choosing interworking equipment is much more difficult.

Intelligence within the network also means that the long distance signalling mechanisms are able to carry the same information that is provided between the user equipment and the local network nodes. This will, in the best case, mean SS7 because other options will likely lose information. For instance, the need to use in-band signalling control for switched 56K service rather than SS7 limits the data rate to 56 Kbps.

The architecture of ISDN, according to ITU-T I.120, will also allow a large variety of configurations. This feature is useful in two ways. First, it is easier to arrive at agreement if more than one method is acceptable (although this may, in itself, cause additional problems for the user). Second, it will more easily allow for growth of the network.

Growth of the network is primarily expected via a commonality of software components. For example, variants on the main signalling protocol (called Q.931) can allow use of other services and bandwidths while retaining the same basic variety of call control information. This commonality of software components is achieved by the use of a *layered protocol architecture.* The Open Systems Interconnection (OSI) model is often used to demonstrate how the software tasks may be divided at particular layers. These layers (explained in greater detail later in this chapter) specify particular points at which the software may be divided.

The layered protocol approach is useful because it allows other OSI-defined protocols to be used with ISDN. It also facilitates development of new protocols based on compatible extensions from existing ISDN-related protocols. Finally, it gives the chance to isolate aspects of the network. For example, a given data protocol should not be directly dependent on the data speed of the transmission network. This would be an aspect of the interactions at the physical layer (to use the ISO nomenclature) or the manipulations of the transmission medium (optical, microwave, electrical). As such, a layered approach allows the network to be extended to new physical platforms without needing changes for higher layers.

THE OSI MODEL

The Open Systems Interconnection (OSI) reference model is the international layered system now used for many new protocol definitions. It is defined by the International Organization for Standardization (ISO). The

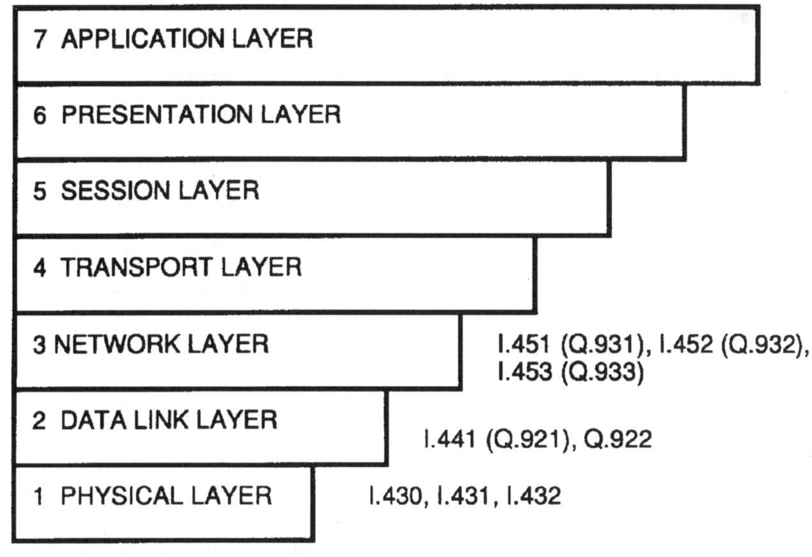

Figure 7.1 OSI model for ISDN.

OSI model provides for seven layers as seen in Figure 7.1. These layers are referred to as the physical, data link, network, transport, session, presentation, and application layers. They are also commonly named by their level number. Thus, the physical layer may be called layer 1. Table 7.2 provides additional detail regarding the OSI model.

Table 7.2 Description of the levels of the OSI model.

Layer	Use
1. Physical	Layer 1 is involved with physical transport needs. Layer 1 is highly dependent on the actual mechanisms used—such as signal voltage for electrical transmissions or signal duration for optical carriers. It will manipulate the physical medium for data transport use. It also involves maintenance of the medium.
2. Data link	Layer 2 works to provide data transfer services, including error detection, retransmission, flow control, and synchronization of messages between the endpoints.

Table 7.2 Continued

Layer	Use
3. Network	Layer 3 is used as a method of providing independence from the actual data transmission. The use of data at this layer may involve signalling; layer 3 will be responsible for connections, ongoing maintenance, and tearing down of the logical connection. The first three layers are often referred to as the "chained layers" within ISDN.
4. Transport	Data-transport services provide end-to-end data integrity.
5. Session	Layer 5 provides connections for the applications. This layer allows for independence from the actual signalling mechanisms but is more dependent on the applications involved.
6. Presentation	Layer 6 provides a standard interface to the application layer and may also provide common services, such as encryption or compression.
7. Application	Layer 7 coordinates requests from the user and the mechanisms used by the protocol and application.

The first three levels (or layers) of the OSI model are often called the "chained layers" because the higher layers are not needed within the network. Thus, some version of the lowest three levels is often found within each network node (layer 3 is needed by nodes involved with gateway functions or routing).

Layers 3 through 5 are concerned with the connection through the network, while the bottom two layers give data transport facilities. The top two layers are involved with the user's needs. It is not uncommon for layers to be brought together within a single software process if the layers are associated with a common purpose. Thus, many PPP programs incorporate layers 3 through 5.

It is also possible, particularly at the higher layers, to leave out a layer that does not serve the needs of the application. For example, there may not be an explicit need for the presentation layer if the application is protocol-specific. Why provide an application interface if only one protocol is to be addressed by the application?

The OSI model provides one main benefit to a protocol stack implementor. It helps to standardize by sections. That is, it is possible to keep independence from the physical medium if all possible media provide the same primitive interface. A physical layer data request from the data link layer uses common parameters. These common parameters are translated by the physical layer into specific hardware and medium requirements. If the physical medium is changed, proper layering ensures that the higher layers will not need to change (as long as the new medium provides the same primitive interface).

OTHER ITU-T I-SERIES RECOMMENDATIONS

We primarily discussed the I.100 series of the ITU-T recommendations so far. These documents are concerned with the overall architecture and potential for growth and change. The remaining documents are more specific in nature. The I.200 "service recommendations" documents are particularly important to the user of services (such as the Internet) over ISDN; they will be discussed in much greater detail in Chapter 9.

The I.300 series corresponds to the use of OSI layer 3. However, rather than protocol-specific documents, these recommendations are concerned with numbering plans, routing, quality of service, and other aspects that must be defined to provide common access methods to the network.

We mentioned earlier that the I.400 series documents often have a second name because these recommendations are involved with specific protocols. Two of the most important ones in ISDN are I.441 (Q.921), which provides the data link layer for signalling within ISDN, and I.451 (Q.931), which provides signalling control for ISDN. These will be covered briefly in this chapter. The I.400 series may be seen as providing the protocol specifications for the software tasks that provide the various OSI layers.

I.500 documents are concerned with interworking. Interworking occurs whenever two or more networks are involved. For example, interconnection of ISDN bearer channels (or B-channels) with the data transport facilities of switched 56 K service is an example of interworking. An international call may also need interworking services to provide mapping of particular parameters or use of standards (such as the digital encoding scheme for speech).

Finally, the I.600 series recommendations are applied to the needs to maintain the network. These may include physical maintenance as well as common methods of subscription information encoding.

BRI AND PRI

The architecture of ISDN is meant to provide a migration toward universal access to integrated services. In particular, this means that people should be able to access services now, without any change in the network access ("telephone line"); it also means that methods must be available for providing broader service to users who need them.

Basic Rate Interface (BRI) supports the first need. BRI can use the same copper twisted-pair wiring that is present in most homes for access to the current analog network. Naturally, being "basic," it provides a minimal set of capacities—particularly in the matter of bandwidth.

Primary Rate Interface (PRI) allows much greater speeds than does BRI. It also requires special wiring that is not normally found at a residential site (although it may be present already at commercial locations).

The primary differences between BRI and PRI reside at the physical layer. BRI provides an interface that gives access to a signalling channel (called the D-channel) that operates at an effective speed of 16 Kbps and two bearer channels (B-channels) that can operate independently at 64 Kbps each or, with physical or logical bonding of the two channels, 128 Kbps aggregate speed. Note that these numbers are for non-compressed data. Many times, advertising for analog modem transport mechanisms will state compressed speeds. Compression algorithms may also be used on ISDN that provide effective speeds much greater than the physical bandwidth.

In North America, PRI provides 24 64 Kbps channels (one of which is normally used for signalling and, thus, designated a D-channel). Elsewhere, PRI usually is defined as 32 64 Kbps channels. One channel is used for physical network needs (framing and synchronization). Once again, one channel is normally used for signalling and is designated as a D-channel. This leaves the remaining 30 channels available for use as bearer channels. Thus, BRI is sometimes known as "2B+D," describing the interface in terms of its channel access; PRI is known as "23B+D" (in North America) or "30B+D" (elsewhere).

Some other differences between BRI and PRI accommodate differences in network needs. Because PRI handles a greater number of B-channels, the allocation and deallocation needs are a little different. It is also possible to group bearer channels into aggregate bandwidths. Six B-channels (for a total of 384 Kbps) are called an H0 channel. With multiple channels set up on a single connection, the manner of passing channel information becomes more complicated than that of BRI.

This is a good time to address the question of why the basic channel capability is 64 Kbps. This rate arises from the needs of a channel for speech capabilities. To digitally "sample" a signal (to get a reading of a value at a particular time), it is necessary to obtain data at twice the highest possible change. Thus, digital encoding requires data to be recorded 8,000 times per second (rounding the normal speech range of 3,500 kHz up to the full 1,000). The next factor is the size of the data. Widely used encoding schemes are able to use only 8 bits per data segment. This 8 bits times 8,000 samples per second yields a 64 Kbps requirement. (Other speech encoding methods now allow use of only 32 Kbps, but this is the historical basis.)

PHYSICAL LAYER USE

The physical layer of ISDN is of considerable importance in understanding how user equipment connects to the network as well as the significance of various endpoint equipment needed and some of the subscription parameters. ITU-T Recommendation I.430 describes the BRI physical layer needs, and Recommendation I.431 explains the PRI physical layer needs. We will concentrate on BRI because this form of ISDN is likely to be in greater demand for Internet use. However, at ISDN provider sites, PRI may end up being a cost-effective mechanism to give sufficient user access.

INTERFACE REFERENCE POINTS

The first thing necessary, before going into physical layer details, is to describe the various "reference points" used by the ITU-T as physical interface locations. Four such reference points are in common use for BRI. These are referred to as the R, S, T, and U reference points. It is not uncommon to refer to the S and T reference points in combination as S/T or just as the S (with T implied).

As can be seen in Figure 7.2, the U reference point is located nearest to the network. This is the physical interface point where the wiring from the local network provider enters into a commercial or residential location. In the United States, this place marks the end of what is provided by the network. In other places in the world, the S/T reference point is included in the physical equipment provided by the network.

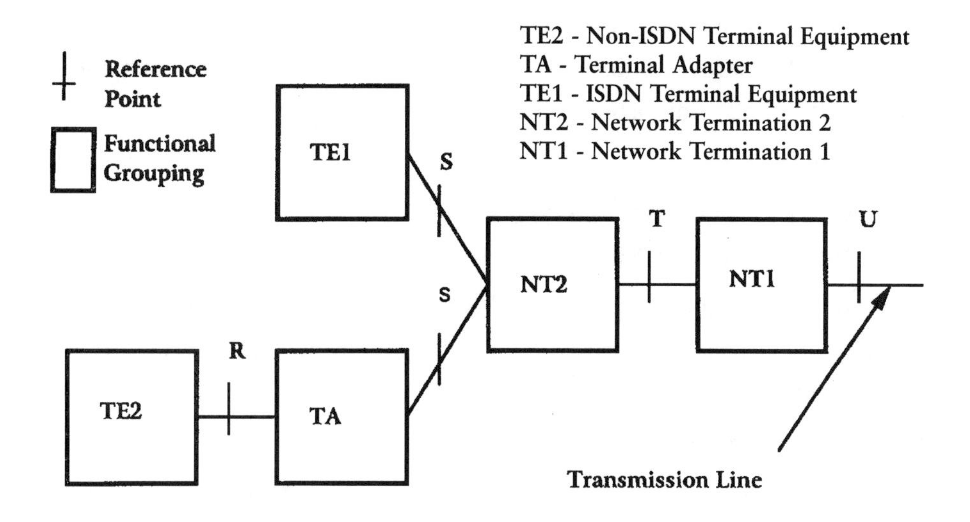

Figure 7.2 ISDN reference points, adapted from ITU-T Recommendation I.411.

This difference is important in the United States. Since the Network Termination-1 (NT-1) spot is marked as the transition point between the U reference point and the S/T reference point, U.S. equipment needs a special device to handle the physical interface translation—known by the transition location label as NT-1 equipment. Some equipment incorporates the U to S/T translation into the user equipment. In such cases, no separate NT-1 device is needed.

The R reference point is also important to the user. The R marks the point between an ISDN terminal adapter (TA) and non-ISDN equipment. This applies in two situations. The first is oriented more toward full signalling applications. Let us say, for example, that you are using a computer and communications package for access to a bulletin board service (BBS). If you are using a serial, RS-232 or equivalent, connection to a modem, you will probably be using the AT command set for access (although this may be hidden by the workings of the communications package). The signalling commands (ATDT, for example) are not the same as those needed for

ISDN, nor are the data in the form that can be sent over the bearer channel easily. The R reference point marks the point where old-styled commands continue to be used by being translated into what is needed by the ISDN equipment.

POINT-TO-POINT VERSUS MULTIPOINT OPERATION

The reference points also provide a degree of flexibility. Because the network enters into the physical location up through the U interface, in the United States, it is possible to provide any type of configuration as long as the U interface standards are kept.

One such configuration is called multipoint or, sometimes, an extended passive bus configuration. This configuration supports the equivalent of extension phones on the old analog lines—with certain distinct differences. These differences arise from the intelligence needed for the general digital protocol. In an analog system, with extension phones, if an incoming call happens then all phones on the line will indicate an incoming call (either by ringing or flashing a line indicator). With ISDN, a particular phone may possibly ignore the signal because the equipment needs indicated in the message are not supported by that device, or because it is addressed to a particular phone, or because all circuits are already busy, among other reasons.

Although some differences exist between a multipoint ISDN situation and an analog set of extension phones, the same types of limitations exist. With a BRI, only two bearer channels can be in *active* use (if the service is supported, one or more calls may be on hold). Thus, if you have four ISDN phones on an extended passive bus, then only two can be active at the same time. This is an improvement over analog situations but imposes a similar limitation. There are also distance limitations and "drain" limitations (that is, you should not use more than a certain number of extension phones due to electrical needs). Most multipoint-supporting networks limit the user to eight devices on the bus.

THE U PHYSICAL LAYER INTERFACE

As mentioned before, the U interface is fully defined only for the United States and some other parts of North America. Other countries consider the U interface to be within the network. For the United States, this interface has been defined by the American National Standards Institute (ANSI) in its T1.601 document. ITU-T Recommendation G.961 gives a partial specification of the U interface for countries other than the United States. A physical layer interface at a reference point is also called a user-network interface (UNI) and is often referred to as such within various documentation.

Digital signals are decoded by means of levels (usually in terms of voltage, for electrical transmission). It is possible to have two levels—and to have more than two levels. In the case of the T1.601 U physical interface specification, four distinct voltage levels are specified. Each of the four levels is translated into one of four values. It is also possible to translate the value by polarity of the signal (positive or negative) as well as the magnitude (amount of voltage). As seen in Table 7.3, a positive voltage level means the first bit is a 1—a negative level means it is a 0. The second bit is interpreted as 1 for low levels and 0 as high levels.

Table 7.3 Two Binary, One Quarternary (2B1Q) Interpretation.

Binary Value	Voltage Level
10	+2.5 volts
11	+0.833 volts
01	-0.833 volts
00	-2.5 volts

Because there are four possible values, the signal is considered to be a quaternary (quat) value rather than binary.

A BRI physical interface is considered to be a 160 Kbps signal frame, which allows for two 64 Kbps bearer channels and the 16 Kbps signalling D-channel. Thus, the data occupies 144 Kbps of the frame. The other 16 Kbps is used for frame maintenance and overhead. This is organized as a repeating frame of 18 bits (9 quats) followed by 12 groups of 8 bits for each B-channel and 2 bits for the D-channel. The frame thus will look something like this:

$S_{18}B1_8B2_8D_2B1_8B2_8D_2$ ... (continued nine more times) ...$B1_8B2_8D_2M_6$

The S bits are used for synchronization, and the M bits are used for maintenance messages. This frame is grouped into what are known as superframes. The synchronization word is used for identification of the first frame of the superframe, and the M bits (with eight frames per superframe, there are a total of 48 M bits per superframe) are used for a variety of maintenance and error-checking purposes.

To keep the 160 Kbps speed, each frame (240 bits long) must occur once every 1.5 ms. The frame will actually be scrambled at transmission and descrambled upon reception. This scrambling of values (discussed in greater detail in other books on ISDN) helps to present a more uniform power distribution (not a net positive or negative voltage level) and can help timing. Scrambling is implemented by replacing specific patterns with other, less troublesome or more easily detectable, patterns. These are then replaced by the original data at the receiver. However, for the purposes of use of the physical interface, only the descrambled output has been presented.

THE S/T BRI PHYSICAL LAYER INTERFACE

Many of the physical criteria for the U interface were imposed on its designers because it was meant as a migratory path to ISDN while using the same physical circuitry currently in place for analog use. Thus, a single twisted pair of wire was considered the basis for physical layer design. This meant that a quaternary coding method was useful to allow greater speeds. It also meant that the wires would need to be used in full-duplex mode (transmission and reception taking place over the same physical lines). Other restraints also caused certain decisions.

In the case of the S/T interface, no such prior constraints were necessary. This is one large advantage in the separation of the reference points. By having the U interface point deal directly with the network (usually over existing physical media), the S/T interface is free to be designed according to equipment needs and protocol usage.

Very little *must* be in common between the U specification and the S interface. The one thing that is necessary is to carry the B-channels and the D-channel at the same data rate. We start, therefore, with 144 Kbps of pure data signal. We then have to allow for balancing of the signal to reduce distortion and power drain. We have to allow bits to keep each frame synchronized (such that the bits are interpreted correctly). Another purpose also arises that is not needed at the U interface—activation. The line between the NT-1 location and the network is normally always active. However, the equipment on the S interface may not always be powered up. You may not desire to have it powered up (and so cause the network to have to pay attention to its idle needs). A final need is to allow for contention among different devices sharing the same bus (a multipoint, or multidrop, situation).

With these various requirements, it was decided to make the BRI at the S interface a 192 Kbps data stream. This allows 48 Kbps (25 percent) as overhead for various purposes. Figure 7.3 shows the bits of a BRI frame structure.

The D-echo channel bit deserves a bit more explanation. This bit is used to detect contention in multipoint situations. When a device has no data link layer frames to transmit, it sends a series of binary 1s on the D-channel. The network side (actually, the NT-1 device if not incorporated directly

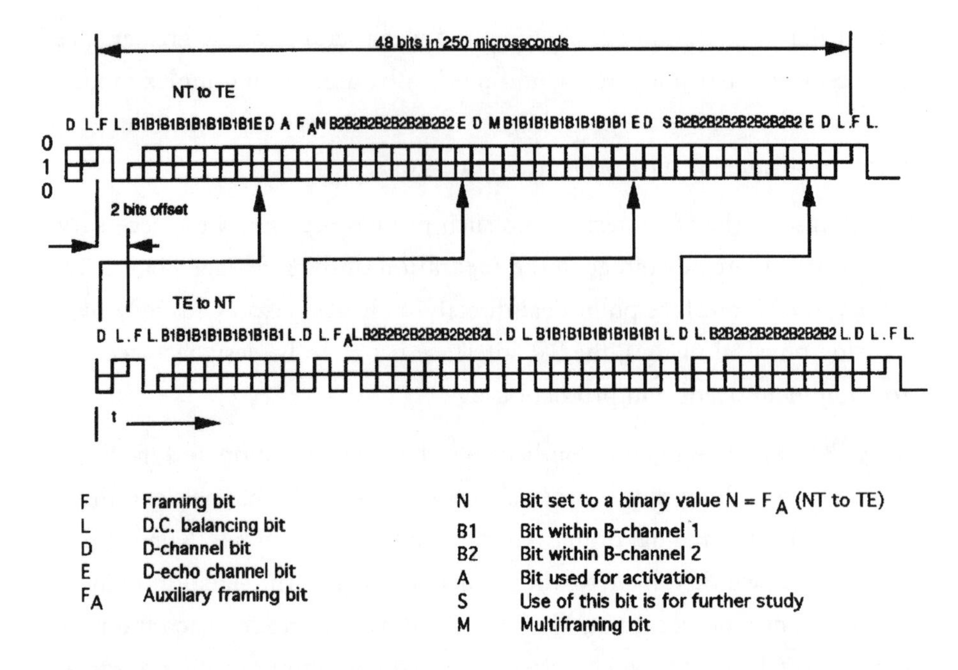

Figure 7.3 BRI physical layer frame structure, adapted from ITU-T Recommendation I.430.

into the equipment) uses the D-echo-channel bit to reflect the value sent to it. Thus, the equipment can monitor the echo bit and, if a series of 1s is detected (the precise number depends on the try number and type of information desired to send), the terminal can assume that the line is idle and start transmitting. At that point, it uses the echo bit to make sure that the NT-1 is receiving what it is transmitting. If not, it must wait until the line appears to be idle again.

Another important design point about the S interface physical layer BRI is that it was decided to use four pairs of wires for the actual bus. This allows a pair each for reception and transmission (each half-duplex). It also allows a pair for possible power transmission. The fourth pair is normally not used currently.

Also, because of the relatively short physical distance covered by the S interface bus, it was much less important to use the higher voltage quaternary encoding scheme. A variation of a binary encoding scheme, called pseudoternary, is used. In this encoding scheme an absent voltage is considered to be of value 1. The value of 0 is represented by alternating positive and negative voltage levels. This alternation keeps the net power use relatively small. Note, however, that this encoding scheme means that long sequences of 1s mean that there will be no voltage-level transitions for that period. This can cause loss of synchronization. Framing bits are used to help ensure that this does not occur.

THE PRI PHYSICAL LAYER INTERFACE

The Primary Rate Interface UNI is usually considered to connect to a T interface reference point because PRI is often used with PBX equipment or other equipment that will split off the channels to other equipment. This may not always be true with increased data use by general equipment.

Two data rates are defined for PRI: 1.544 Mbps and 2.048 Mbps. The 1.544 Mbps is based on the North American DS-1 transmission structure, which is used with the T1 transmission network. The 2.048 Mbps is based on the European E1 transmission structure, as defined in ITU-T Recommendation G.704.

The amount of overhead within a PRI link is considerably smaller than that for BRI. It amounts to 1 bit out of 193 for the T1 version and 8 bits out of 256 for the E1 structure. Each frame is transmitted within 125 μsec. This allows each frame to carry the appropriate total data rate [$8000 \times (193 \text{ bits}/125\ \mu\text{sec}) = 1.544$ Mbps for T1, $8000 \times (256 \text{ bits}/125\ \mu\text{sec}) = 2.048$ Mbps for E1].

The frame for T1 transmission consists of a framing bit followed by eight bits each for the 24 channels allowed in the North American PRI. These 193 bits make up one frame. These are grouped together into a multiframe composed of 24 193-bit frames. These 24 framing bits are used for synchronization purposes and for error checking on the framing bits. Error detection on individual channels is the responsibility of the protocol used within the channel.

The frame structure for E1 transmission divides the 256 bits of the frame into 32 equal channels of 8 bits each. The first channel is considered to be a framing channel. This fulfills the same general purposes as does the T1 frame structure. The E1 structure, however, does make greater allowance for international differences in transmission—thus allocating a larger number of overhead bits for this use.

PHYSICAL LAYER PRIMITIVES

Each of the OSI layers must be able to communicate with the other layers. The physical layer must be able to work with the data link layer, and vice versa. This is done via primitives. There are primarily two entities with which the physical layer must work: the data link layer and a cross-layer (sometimes referred to as a plane layer, indicating that it interacts "horizontally" through more than one layer) entity sometimes referred to as the management entity. It is also referred to as the C-plane, or control plane.

The ITU-T recommendations give four general types of primitives that are to be used by a layer. These are requests, indications, responses, and confirmations. A request is a primitive indicating needed service by a higher layer to a lower layer (the management layer is conceptually higher than all of the regular layers). Thus, the data link layer makes requests of the physical layer. A lower layer may respond (if it is not a synchronous request, such as a function call) to a request with a confirmation. Requests and confirmations are loosely bound together.

An indication is a notification, or service message, from a lower layer to a higher layer. It may be replied to by a response primitive. The name of a primitive is thus composed of an interlayer identifier, primitive identifier, and direction/use notification. A physical layer request for data transmission is a PH_DATA_RQ primitive.

The ITU-T defines only request and indication primitives for the physical layer because all such messages are considered to be autonomous and unable to be refused. Three primitive groups are defined for interactions between the data link layer and the physical layer. These are the PH_DATA_RQ and PH_DATA_IN primitives, which request the physical layer to transmit data and received information from the physical layer that is to be passed to the data link layer. There are also the PH_ACTIVATE_RQ and PH_ACTIVATE_IN primitives, which allow the data link layer to start activation of the physical layer and an indication that the physical layer is now ready for data transmission and reception. Note that, for the U interface and for PRI, the physical line is *always* supposed to be active and, therefore, the PH_ACTIVATE_RQ primitive is not used in those situations. Finally, we have the opposite of the PH_ACTIVATE_IN primitive—the PH_DEACTIVATE_IN, which indicates the physical layer may not be used.

The management entity is used as a system mechanism to keep track of system status. Primitives that are used by it tend to be status information and are often only indications. Four indications are defined for the physical layer and one request. The request is an MPH_ACTIVATE_RQ primitive (note the MPH, which indicates a management entity to physical layer primitive)—used only for S/T BRI (and optional even there). The indications include a parallel MPH_ACTIVATE_IN and MPH_DEACTIVATE_IN primitive plus an MPH_ERROR_IN primitive, which allows statistical information gathering for problems at the physical layer. The final indication is the MPH_INFORMATION_IN primitive, which is used for S/T BRIs that make use of the third wiring pair for power (this is optional).

These primitives are not really necessary to know to make use of ISDN equipment, but it is useful to be aware of just how the layers communicate to make a fully integrated, functioning system.

DATA LINK LAYER FOR SIGNALLING

The physical layer takes care of framing for the physical transmission medium. In other words, it makes sure that a 1 bit is treated as such and that the data part of the medium aligns in the same manner for reception as created at transmission time. However, it does not do anything for the actual data.

There are two main possibilities for the data being carried over a transmission line. They are continuous, or they are sporadic. Examples of continuous data include television transmissions carried over coaxial cable or via satellite. Another would be radio programs or speech (though few people talk continuously). A large number of data uses are not continuous. There is a beginning of useful data and an end. When data are not continuous, a protocol that delimits the data is needed. This protocol often incorporates some type of error checking with possible retransmission of erroneous, or out-of-sequence, data.

The data link layer is responsible for this. Normally, continuous data are handled by specialized hardware devices or dedicated software modules, and they do not fit into the OSI model. The high-level data link control (HDLC) family of data link protocols is often used for layer 2 data link procedures.

HDLC DATA LINK FAMILY

This family of protocols allows many options within a common framework. The basic commonality exists in the ways that the frames are delimited and the general methods of addressing and controlling information. If a stream of data is said to begin and end, then there must also be times when it is idle (no valid data in the stream). This period is called the idle time. Idle periods are noted by the use of idle markers, which are often expressed by flags or marks. A flag in HDLC is composed of a set pattern, which is a 0 followed by six 1 bits, ending with a 0. The first non-flag byte is considered to be the first byte of the real data. This real data stream is ended by the next flag character. This method is used by link-layer access protocol for D-channel (LAPD) that is used by the signalling channel in ISDN. For some other protocols (used on the B-channels), a "mark idle" may be used—the mark is a stream of continuous 1s.

In any system that uses a particular pattern to express a meaning, there will also need to be an escape method to allow that pattern to be used within the other mode of data. This is true for HDLC also. If a hexadecimal 0x7E (01111110) needs to be part of the data, it must be transformed before transmission and restored upon reception (similar to scrambling at the physical layer). This is done by 0-bit insertion. If data are being transmitted and a sixth consecutive 1 is desired, a 0 will be inserted after the fifth 1. Thus, the pattern 01111110 will be transmitted as nine bits—011111010. On receipt, a 0 that follows five 1s will be deleted. This restores the original data.

Within the valid data frame, HDLC allows for a sequence of bytes. These begin with one, or more, address bytes. Address bytes allow routing of the data at the data link level. The next field, called the control field, determines the actions of the protocol upon receipt of the frame. The next field is optional, depending on the control field. It is the data field and, if present and the rest of the frame is correct, it contains information that will be forwarded to layer 3 of the protocol stack. The frame is normally ended with a frame check sequence (FCS). For LAPD, this FCS is formed by a 16-bit cyclic redundancy check (CRC-16), which allows some data integrity checking.

HDLC truly is a family of protocols. It allows a variety of options such that the protocol may be balanced (both sides can send and receive the same messages) or unbalanced (one side is primary and can send commands that are illegal for the other side to send). It may be full-duplex (asynchronous) or half-duplex (synchronous). Examples of various HDLC protocols include SDLC, LAPD, LAPB (X.25 layer 2), LAPM (link access protocol for modems), V.120 layer 2, and so forth. Some protocols are very similar. Others diverge at a low-enough level (such as balanced versus unbalanced) that the rest of the protocol is considerably different. They all are delimited frames that contain address fields, control fields, and (if appropriate) data fields.

LAPD ADDRESSING

The first two bytes of an ITU-T Recommendation Q.921 (LAPD) frame are considered to be the address field. As can be seen in Figure 7.4, there are three subfields of the address. These are the command/response (C/R) subfield, the service access point identifier (SAPI), and the terminal endpoint identifier (TEI). The low-order bit of the first byte is a 0 which is a common ITU-T method of indicating the continuation of a field. Similarly, the 1 in the low-order bit of the second byte indicates the end of the field.

LAPD is an asynchronous, balanced, HDLC protocol. It is also considered to be an extended HDLC protocol, based on its use of modulo 128 sequencing numbers for acknowledged transfer mode. In the case of a bal-

8	7	6	5	4	3	2	1	Byte
SAPI						C/R	EA 0	2
TEI							EA 1	3

EA = Address field extension bit
C/R = Command/response field bit
SAPI = Service access point identifier
TEI = Terminal endpoint identifier

Figure 7.4 LAPD header address field, adapted from ITU-T Recommendation Q.921.

anced mode, commands and responses will be available from both sides. It would be possible to have extra bits reserved in the control field to indicate the category of control information. It was decided by the ITU-T to not do this.

The C/R subfield allows designation of the role that is being played by the peer side of the protocol. In the case of LAPD, terminal equipment will play one role and the network will play the other role. The terminal equipment side sets this bit to 0 to indicate commands and 1 to indicate responses. The network side uses 1 to indicate commands and 0 to indicate responses. This has a historical basis from the LAPB of X.25 layer 2. In many ways, it would be much easier to have a fixed configuration for all endpoints (this is done for later LAPD derivatives such as V.120).

TEIs

The terminal endpoint identifier (TEI) is something to which an ISDN user may have indirect access. Three value ranges apply to LAPD. The values from 0 through 63 are considered to be *fixed*. These are set up at subscription time with the network and can be used only for point-to-point configurations (including PRI). The value of 127 is considered to be a broadcast value. This is used when the specific equipment being addressed is unknown by the network.

The range of values from 64 through 126 are considered to be automatic. They are requested by the terminal equipment and assigned by the network. Each terminal, using automatic TEIs, will request a TEI before communicating with the network. The ITU-T recommendations state how TEIs are to be managed and negotiated, but they are not specific as to how many are to be used. In the case of two specification variants (Northern Telecom DMS-100 and Bellcore U.S. National ISDN-1) used in the United States, a separate TEI is used for each B-channel that is available.

TEI negotiation takes place over the broadcast link (TEI 127) between the management entities of the network (this is designated by a specific SAPI, which will be discussed in the next section). The terminal equipment sends an unacknowledged information frame over the broadcast management link requesting a TEI. The network responds with a TEI assignment (in the best case). Other messages can also occur on the data link, including TEI verification and removal.

Once a TEI has been obtained from the network, specific layer 3 packets may be exchanged. The TEI will be unique to each user equipment device on a multipoint drop (or for a point-to-point situation where only one device is present). This allows messages to be ignored by non-appropriate devices on a multipoint configuration and verifies identity in point-to-point situations.

SAPIs

The service access point identifier (SAPI) is employed to indicate the use of a particular data link. The SAPI is needed to route the message to the appropriate layer or entity for further processing. Three SAPI values were originally reserved by the ITU-T for specific uses. Other values are now being reserved for continued architectural expansion.

The three original reserved values were 0 (used for signalling information), 16 (used for X.25 packet mode over the D-channel) and 63 (used for management procedures, such as TEI negotiation). The value of 1 has now been reserved for additional packet-mode signalling in preparation for integration of older systems with broadband applications. Other ranges may end up being used for switched virtual connections (SVCs) for packet services such as frame relay. (Current recommendations call for this, but network evolution has not yet reached this point of use.)

The combination of TEI and SAPI values determine routing. A broadcast management message (TEI 127, SAPI 63) indicates Q.921 management procedures. This currently means only TEI management (but it may be expanded in the future without need to change the architecture). A broadcast signalling message (TEI 127, SAPI 0) may be used by the network for general information distribution or for indication of an incoming call on a multipoint situation.

The use of the signalling TEI combined with signalling use (TEI 0 - 126, SAPI 0) indicates that routing of data to the Q.931 (network layer signalling module) is appropriate. Similarly, a packet data link (TEI 0 - 126, SAPI 16) is routed to a layer 3 packet handler (normally X.25 layer 3). The use of non-broadcast management (TEI 0 - 126, SAPI 63) is currently not defined (but it may be in the future).

Note that the combined fields of the TEI and SAPI are commonly referred to as the data link control identifier (DLCI). Each of the specific uses and routings discussed above is associated with a DLCI (combined address subfield). In other LAPD-derived protocols, such as ITU-T Recommendation V.120 data link layer, this combined field may overlay the TEI and SAPI fields and be used in a protocol-unique manner. The reserved values of the SAPI field help facilitate this multiprotocol use of the address field.

LAPD PROTOCOL BASICS

LAPD is described as a state-machine (as is the physical layer, I.430, and many other protocols) and is normally implemented as such. A state-machine consists of three components: the state, an event, and a set of responses. The state is a description that indicates history of events and current configuration parameters. There will always be an initial state—the state where no external event has yet occurred. Within a given state, a set of responses will be executed when an event occurs. These responses may include the changing of the state—which causes future events to be handled in a potentially different manner.

LAPD has eight different states. The states considered states 1 through 3 are TEI-unassigned states. These states allow transmission and reception of broadcast messages but do not allow specific data link operation. State 4 is designated the TEI-assigned state. At this point, specific data links may be addressed but only as unacknowledged operations. States 5 and 6 are transitory states used during establishment or teardown of acknowledged operation. State 7 is the normal multiple frame established state, and it is the state during which most LAPD activity is expected to occur. State 8 is a timer recovery state, which allows reception of further data but holds off on transmission until acknowledgment of previous data has occurred.

It is possible that, for PRI situations or other fixed TEI implementations, only states 4 through 8 will be implemented. In such an implementation, state 4 will be the initial state. It is also possible that acknowledged operation will never be needed by the higher layer protocols. In such a situation, only state 4 is used. This indicates the versatility of the LAPD protocol and the variety of specific uses for which it may be used.

In ISDN signalling (except for PRI), all eight states are used. The states will be used in two ways. The first is called unacknowledged mode. Broadcast messages may occur in any state. Data link specific messages will be allowed in states 4 through 8. The second is called acknowledged mode, which is allowed only in states 7 and (for reception) 8. Signalling information is usually initiated by the network with an unacknowledged information frame (allowing for the possibility that TEIs have not yet been initiated) on an incoming call direction. All signalling traffic from the user equipment will occur with an acknowledged information frame.

LAPD CONTROL FIELD

The control frame of LAPD, as seen in Figure 7.5, consists of one or two bytes. They will be interpreted in conjunction with the value of the C/R bit contained in the address field. The length is determined by examination of the low-order two bits of the first (perhaps only) byte.

Extended addressing Control field bits	8	7	6	5	4	3	2	1	Byte
I format	N(S)							0	4
	N(R)							P	5
S format	X	X	X	X	S	S	0	1	4
	N(R)							P/F	5
U format	M	M	M	P/F	M	M	1	1	4

N(S)	Transmitter send sequence number	M	Modifier function bit
N(R)	Transmitter receive sequence number	P/F	Poll bit when issued as a command. Final bit when issued as a response.
S	Supervisory function bit	X	Reserved and set to 0

Figure 7.5 LAPD header control field, adapted from ITU-T Recommendation A.921.

INFORMATION FRAMES

A value of 0 in the low-order bit will indicate an acknowledged Information frame. The remaining parts of the two control bytes will be used to indicate sequencing information and the poll/final (P/F) bit.

The P/F bit is used as a request for or response to a need for explicit acknowledgment. It is considered a poll bit for use within commands and a final bit when used within a response. (A poll bit set to 0 indicates that no specific acknowledgment is mandated by the protocol. A final bit set to 0 in response to a command with poll bit set to 1 will not achieve the desired acknowledgment. A final bit set to 1 without a pending poll bit in the received command will be considered an error by the protocol.) In LAPD, only information commands are legal, and thus the P/F bit is designated only as a poll bit. (Note that, in V.120 layer 2, information responses are also legal.)

LAPD makes use of extended sequencing information—modulo 128. This means that each frame may have a sequencing number in the range of 0 through 127 (with the value of 0 following a value of 127). HDLC protocols also allow for a non-extended sequencing of modulo 8 that permits use of a single byte for all control fields (used by LAPB for X.25 layer 2). When initial conditions are established, all sequence variables will be set to a value of 0. Thus, any information sent will start with sequence value 0, and the next frame will have sequence value 1, through value 127 and then back to 0. The receiving side expects the acknowledge information frames to arrive in order. Thus, if sequence number 0 is followed by sequence number 2, it indicates that a frame is lost; retransmission and recovery will be initiated.

The other sequencing number field in the information frame is used for acknowledgment of received frames. Thus, the first value is referred to as N(S) (for sending) and the other is referred to as N(R) (for receiving). It is possible for the receiving end to acknowledge more than one frame at a time. This is a side effect of windowing that will be discussed shortly.

Supervisory Frames

Supervisory frames are denoted by the use of 01 in the low-order two bits of the first byte of the control field. Bits three and four (starting numbering at bit 1 for the lowest-order bit) are used to distinguish between four possible supervisory commands. In LAPD, only three values are defined. These are the receiver ready (RR), receiver not ready (RNR), and reject (REJ) frames. Each of these supervisory frames includes an N(R); the primary purpose of these frames is to acknowledge activity of, and specific data transfers from, the data link.

An RR indicates normal response with the local protocol still able to receive data frames. An RNR indicates that the local entity is busy and unable to receive further data. A REJ frame indicates that an error has occurred on the data link but still uses the N(R) to acknowledge data frames up to the frame in error.

Each supervisory frame can be used as a command or as a response. A supervisory frame used as a response will echo the value of the received supervisory (or other frame type) frame's poll bit within its final bit. A supervisory frame used as a command will normally have its poll bit set to 1. This allows multiple use of supervisory frames. It may be used as a passive command/response (allowing quick data transfer). It may be used as a polling method to make sure that the data link is still active.

Let's examine the REJ frame a bit more. Assume that frames 0 through 5 were transmitted by the peer. The receiver get frames 0 through 3 and then 5. The REJ frame will thus set the N(R) to 4, indicating that the next valid sequence number to be received is frame number 4. Since the transmitting side has already sent frame 4, the receipt of this message also indicates that it must retransmit starting with frame 4. The sequence number serves two purposes: it acknowledges frames 0 through 3, and it indicates to the transmitter where data retransmission must start.

Unnumbered Frames

Unnumbered frames are recognized by the value 11 in their low-order bits. As indicated by the name, unnumbered frames do not contain sequencing information and, thus, have the entire control information within a single byte. Since the second byte is not used, the poll/final bit must be incorporated into the initial control byte. Most unnumbered frames are used only as a command or as a response. These frames are used to change protocol states, transfer unacknowledged data, verify the activity of the link, specify protocol information, or determine protocol parameter exchanges.

The two commands within the first category are the set asynchronous balanced mode extended (SABME) command and the disconnect (DISC) command. The SABME command will initiate transfer from an unacknowledged state to a multiple frame established state (state 7 or 8). It may also be used to reinitialize the data link in case of protocol errors. The DISC command is used to change the state back to the unacknowledged state from the multiple frame established state. This command is rarely used within a normal ISDN system.

The other frame in the first category is a disconnected mode (DM) response frame. This is sent as a response to a command; it is used only within multiple frame established states, when the receiver is not in multiple frame established state. Thus, the DM response frame is normally encountered only in error conditions.

The unnumbered information (UI) command frame is used to send unacknowledged data. It is used within ISDN signalling only for management data by the terminal equipment, but it may be used by the network for various purposes. A UI frame does not contain a sequencing number and is normally not explicitly acknowledged. It is the responsibility of higher layers to make sure that the data have been transferred correctly. It is very fast, however, and is, in essence, the way that frame relay transfers data.

The unnumbered acknowledgment (UA) response frame is normally used within unacknowledged states as a response to an unnumbered command that requires state changing (SABME or DISC).

The frame reject (FRMR) response frame is used when the received frame indicates a protocol violation error (rather than simply a lost frame). It will use the data section of the frame to indicate the specific reasons for the rejection and an echoing of the erroneous bytes of that frame. Within LAPD, the FRMR response frame is normally generated only by the network (in X.25 layer 2, LAPB, it may be generated by the terminal as well). An FRMR is often preceded by a SABME, but, in either case, it will indicate a need to reinitialize the link, and (because it is a protocol violation) it may need to be reported to the management entity for tracking of the problem. Note that a corrupted frame that is not detectable by the FCS may cause this problem within a well-implemented system.

The final unnumbered frames defined by Q.921 are exchange identification (XID) commands and responses. The use of these is not fully defined by ITU-T Recommendation Q.921. However, they are meant to be used as protocol parameter negotiation mechanisms. For example, if modulo 8 (rather than 128) is desired, the XID frame could communicate this need. It could also negotiate the window size, as discussed in the following section.

WINDOWING

Frame acknowledgment must occur in order. However, it does not have to be consecutive. This involves what is called a window. The window size, designated by the variable *k* within Q.921, indicates the total number of acknowledged information frames that may be outstanding (unacknowledged) at one time. A number of default values are established for this window size. For signalling purposes in BRI (SAPI 0), the window size is 1. A window size of 1 actually negates the concept of windows. However, for other uses (SAPI not equal to 0), the default BRI window size is 3. For PRI, the default window size is 7.

A window size delimits the frame sequencing numbers that may be valid in transmission. Thus, if the window size is 3, then frames number 0, 1, and 2 may be outstanding. It is also possible that the outstanding frame numbers may be 120, 121, and 122. Two internal variables keep track of the sequence numbers in this situation. One keeps track of the next sequence number to be acknowledged. The other keeps track of the next sequence number to be used in acknowledged information transfer. These are known as V(A) and V(S). The value of N(S) in a transmitted I frame will be equal to V(S) (and then V(S) will be incremented).

Windowing is particularly important in situations where the transmission delay is significant. This might occur in satellite transmission situations or in data situations where the data must be internetworked. The concept of windowing also exists in other layers where acknowledgment of data occurs.

PACKET/FRAME SIZES

For most switches, the maximum frame size for a D-channel message will be 260. This value allows for 4 bytes of LAPD header information (the address and control fields), plus 256 bytes of data that can be used between layer 2 and layer 3. For early INS-Net 64 (Japan) implementations, this value was 132.

DATA LINK LAYER PRIMITIVES

As was true for the physical layer, it is necessary for the data link layer to communicate with other layers. The primitives dealing with the physical layer have already been discussed. Like all ITU-T protocol layer recommendations, the primitives fall into the four categories of requests, confirmations, indications, and responses.

Four primitive types are used between the data link layer and the network layer. These are the DL_ESTABLISH_RQ (also indication and confirmation) primitive, DL_RELEASE_RQ (also indication and confirmation) primitive, DL_DATA_RQ and DL_UNIT_DATA_RQ (and upward-going indication) primitives.

The DL_ESTABLISH_RQ primitive is used to cause a transition to multiple frame established state. Note that the data link layer may not be in a TEI-assigned state when such a request arrives. This explains the previously unexplained states of 2 and 3 that exist for LAPD (allowing for a desire for multiple frame established state before the TEI assignment has occurred). This change of state may occur from the locally initiated DL_ESTABLISH_RQ primitive, in which case a successful response will be a DL_ESTABLISH_CF, or it may be initiated by the other side, which will cause a DL_ESTABLISH_IN primitive to be generated.

The DL_RELEASE_RQ (and indication and confirmation) primitives are used in a similar manner. However, in these cases the information transferred is that the data link is no longer in multiple frame established state. As mentioned earlier, in conjunction with the DISC command frame, it is rare that the higher layers will intentionally cause this. Thus, the DL_RELEASE_IN primitive is the most likely primitive—probably originating from line transmission problems (indicated by a PH_DI or from continued data link problems).

The DL_DATA_RQ and DL_DATA_IN primitives are used for acknowledged information transfer to and from the network layer. In a similar fashion, the DL_UNIT_DATA_RQ and DL_UNIT_DATA_IN primitives are used for unacknowledged information transfer. It should be noted that the DL_ primitives differ from the physical layer, PH_, primitives in one important fact. There is normally a single line involved with the physical layer. There will be different data links supported on the same channel. Thus, primitives must carry identification information as well as parameters that allow for the command to be properly executed (such as the location and length of data).

ISDN SIGNALLING PROTOCOL FOR THE NETWORK LAYER

The network layer protocol for ISDN signalling is defined in ITU-T Recommendation Q.931. This protocol is passed between the network layer and the data link layer using the DL_ primitives. The DL_ESTABLISH_RQ primitive may be generated by the network layer or the system during initialization. Generally, in North America, the link will need to be established as soon as the terminal is powered up. In other countries, this is normally done only when there is a need to communicate information (an outgoing or incoming call is desired). The network will generate the release of the link when it has determined the line is no longer in immediate use.

The DL_EST_IN and DL_REL_IN primitives are used by the network layer to determine when the data link is available for use. The DL_DATA_RQ and DL_UNIT_DATA_RQ primitives are used for data requests as appropriate (along with corresponding indications).

Q.931 PROTOCOL BASICS

Q.931 is also defined as a state-machine. The states are defined on a per-call basis. Each potential connection will start in the NULL state (state 0) and will try to reach the ACTIVE state (state 10). States 1 through 9 are basically used during the establishment phase of a call (incoming or outgoing). States 11 through 19 are used during the teardown of a call or during the process of suspension/resumption of a call. States above 19 are used for special purposes and cannot be easily categorized.

The network and user sides each have their own set of states. However, most states roughly correspond to the other side. Events are composed of data link primitives, the contents of the (up to 256 byte) data field passed from the data link layer, or primitives from higher layers or the management entity. Responses, once again, may involve a change of state and messages sent by the network layer (to higher layers or to the peer entity).

One mention should be made here (which is also applicable to the data link layer) of timers. Timers exist to make sure that information is properly acknowledged in a timely fashion. This is particularly important at the network layer since it does not directly control the retransmission of the data frames. It will normally *regenerate* any message frames that must be retransmitted. This is different from the data link layer, which normally stores the data received from the higher layer until acknowledgment is received. This regeneration means that information must be retained within the protocol state information to allow such recreation of the appropriate message.

Q.931 MESSAGE CONTENTS

The message contents of a Q.931 message consists of a protocol header, call reference value (CRV) information, a message type, and information elements that are associated with the particular message.

For Q.931, the protocol header value is a byte containing the value of 8. Some pre-ISDN data protocols, such as Germany's 1TR6, make use of this protocol header byte for their own message contents. It can also be used to allow re-routing of messages to other signalling entities such as V.120 (which can use a subset of Q.931 messages for in-band call control).

The CRV is very important within ISDN signalling. This value is assigned by the originator of each call. The upper bit of this value is used to indicate the direction of origination. (This allows each side to allocate CRVs independently—reducing the potential of conflict in assignment.) The first byte of this field (the second byte of the Q.931 message) is actually the CRV length. A length of 0 is used to indicate that this is a general network message (for some supplementary services, a special "dummy" CRV is used to indicate the same type of information).

Most specifications call for a single byte CRV—allowing 127 CRVs to be allocated by each side. However, the protocol allows for multiple bytes (and the Northern Telecom DMS-100 specification does allow for two byte CRVs, although most networks only initiate single byte CRVs). When multiple bytes are allowed within a field, Q.931 uses the "extension bit" within the bytes (usually the high-order bit within the byte) to indicate the last byte of a field—with a 1 indicating the last byte.

The next byte following the CRV length and (if present) CRV contents is the message type. The message type and information elements are interpreted within what is called a *codeset.* Codeset 0 is considered to be the base codeset. ITU-T Recommendation Q.931 allows for direct escape into a national codeset by use of a special message type. Normally, however, codesets are only shifted (or changed; shifted is a historical reference to the "shift" operation of manual typewriters) within the information elements (IEs) that are associated with a particular message type.

MESSAGE TYPES

Message types fall into five categories, according to the ITU-T Recommendation. One special message type, of value 0, indicates that the following bytes are to be interpreted according to national needs. The others fall into the categories of call establishment (upper three bits value 000), call information (upper three bits value 001), call clearing (upper three bits value 010), and miscellaneous messages (upper three bits value 011).

We will concentrate on call establishment and call clearing message types because these are the types of messages most often used on ISDN lines. Call establishment messages include ALERTING, CALL PROCEEDING, CONNECT, CONNECT ACKNOWLEDGE, PROGRESS, SETUP, and SETUP ACKNOWLEDGE. (Message types are usually referred to in all capital letters to distinguish between a sort of a message and a specific hex value equivalency for protocol reasons.)

One way of looking at these message types is to see what their analog network equivalents would be. Each message type fills a certain need of the system. Because analog systems have been around a long time (relatively speaking) and have been the most intensely studied, it is to be expected that many digital signalling messages will follow analog mechanisms. SETUP is the same as dialing (or using touch-tone) digits to initiate a call. SETUP_ACK is roughly the same as a dial tone. It acknowledges the desire to place a call, but it does not have enough information to have the network try to establish the call. CALL_PROC and PROGRESS both allow for feedback to the originator of a call as to status (similar to ringback on analog phones). ALERT is received by the destination and is the digital equivalent to a ringer being started. Finally, CONNECT is the same as picking up the receiver to answer a call, and CONN_ACK is the removal of ringback.

Call clearing messages include DISCONNECT, RELEASE, RELEASE COMPLETE, RESTART, and RESTART ACKNOWLEDGE. RESTART and RESTART_ACK have no analog equivalents. They are oriented toward the equipment (or network) as a whole and are meant to request (and acknowledge) the resetting of the entire system software back to an initial state. A DISCONNECT message is the same as hanging up a receiver. RELEASE and RELEASE_COMPLETE are difficult to find direct analogies for, but they are similar to a "dead signal" (no dial tone, ringback, or voice) on an analog line. RELEASE asks for the call to be terminated. RELEASE_COMPLETE indicates that the call *has* been terminated. The sending or receiving of a RELEASE_COMPLETE message type will cause (or indicate) the deallocation of the CRV and, thus, no additional messages for this call are valid.

Message types will be normally used as part of sequences. Examples of these sequences will be explained, in detail, in the next chapter. However, there will always be normal scenarios and error scenarios. A normal scenario is one in which everything works correctly and a call is correctly placed or disconnected. An error scenario has problems that must be handled.

Information Elements

Following a message type, there will usually be a series of IEs. An information element may be a single byte information element (indicated by a 1 in the high-order bit of the byte), or it may be a variable-length information element (indicated by a 0 in the high-order bit). If it is variable length, the next byte will contain the number of bytes associated with the IE. Thus, one has IE code, length x, and then x bytes of information associated with the IE.

Each message type will have a number of IEs associated with it. Some will be listed as mandatory—they must be present for the message type to be considered valid. Some are optional—they may be present or not. A third set (usually described by notes or specific instructions in the documentation) will be optional under certain circumstances and mandatory otherwise. This third set is the hardest to be certain of proper implementation because each optional byte must be analyzed, and properly accepted or rejected, by the network or test equipment. If this is not done, the specification will be inadvertently altered.

Two of the most important IEs are the bearer capability IE and the channel identification IE. These tell what the bearer channel is to be used for, and which bearer channel (or channels, in the case of PRI or multiple channel connections) is to be used. Other IEs are primarily informational. They may be very important to an application for determining correct procedure, but they are not required to set up the link.

SPIDS

The service profile identifier (SPID) was originally designed as a method to distinguish between different types of equipment put on a multipoint line. The SPID was used to determine the exact services that could be supported by the equipment. Often the SPID is a concatenation of terminal type information with the dialed number associated with the line (or channel).

However, original designs do not always continue unchanged. Currently, SPIDs are used in point-to-point as well as multipoint situations. They are still used as identifications of equipment, but they may be superfluous in that only one piece of equipment is on the line. The reason for this change in architecture was probably twofold. One was to allow different equipment to be used on the point-to-point connection. For example, an ISDN TA is put on the line. Its SPID may be different than that of a Group 4 fax device. Second, it allows a greater uniformity for setting up ISDN equipment with a switch. It is no longer necessary to decide whether the SPID is required—it is used on almost all connections using non-fixed TEIs. The SPID is used only for North American ISDN variants.

The SPID is sent to the network as soon as the TEI is obtained. The network should respond to the SPID with a two-byte IE known as the equipment identifier (EID). This EID may be broken into two components—the user service identifier (USID) and terminal identifier (TID).

Originally, the network would (after SPID/EID exchange) always put the EID into any broadcast message meant for a particular terminal. This also has changed. The EID is not always used—the dialed number is used as a distinguishing characteristic.

SUPPLEMENTARY SERVICES

Supplementary Services for ISDN are covered, in basics, by ITU-T Recommendation Q.932. This document explains how such services may be implemented—the basic procedures. The exact services, and coding of such messages, are left to the switch-specific documents. Thus, EuroISDN will have a set of supplementary services, and AT&T 5ESS will have a set of supplementary services. The procedures will follow the ITU-T Q.932 recommendations, but the exact coding may be different.

Supplementary services provide the types of features found in custom calling for analog lines. Thus, there may be the ability to place a call on hold—or set up a conference call—or transfer the call to another line. It may also include multiple call appearances, where multiple connections may be in use (although only two bearer channels may be actively carrying data).

HIGHER-LAYER ACCESS

As is true of all layers, there must be methods of communicating information and requests across the layer boundaries. In the case of layers above the network layer, the ITU-T recommendations only give partial solutions. In order to be consistent, it is useful to retain the concept of requests, confirmations, indications, and responses. The exact primitives, however, depend on how the equipment is designed. It would be possible to have no upper layer primitives if the only purpose of a device was to bring up a voice connection to a fixed address as soon as the device was powered up.

Generally, this very simple case is not enough to provide a general environment for use of ISDN. An N_CONN_RQ, N_CONN_CF, N_CONN_IN, and N_CONN_RS set of primitives is very useful to allow incoming and outgoing call setups. Similarly, the N_DISC_RQ, N_DISC_CF, N_DISC_IN, and (optionally) N_DISC_RS primitives are useful to tear down a call. If data protocols are supported within the software system,

N_DATA_RQ and N_DATA_IN primitives may be very useful. However, if bearer channel use is done in a separate software module, it may be done within that protocol stack in its own way.

Each ISDN device has its own needs, which are based on the switch variant being used, the purpose of the equipment, and the types of services provided. The next chapter will help you understand how ISDN works for particular situations by detailing the primitives involved in specific scenarios.

CHAPTER

8

ISDN SIGNALLING USE

Most of the activity considered to be ISDN concerns the use of ITU-T Recommendation Q.931 (I.451). The protocols used at the physical layer and data link layer are very important in integrating user equipment with the network. However, they are general tools, and the data link layer in particular may be used for a variety of purposes not directly related to ISDN.

Thus, there is need for a general understanding of the message types and their uses for Q.931 within the network and user interface. One of the most straightforward ways to do this is to give examples of scenarios that will be used within ISDN.

REVIEW OF OTHER LAYER SUPPORT FOR Q.931

Within the physical layer (I.430 for BRI, I.431 for PRI), the primary function is to provide a reliable physical link between the equipment and the network. This is done by specific physical transmission, and framing, specifications. The physical frame is always divided into separate TDM channels. In the case of BRI, these channels are divided unequally. The signalling channel, or D-channel, is 16 Kbps. The two bearer channels, or B-channels, are each 64 Kbps. This allows 8 Kbps of data per B-channel for signalling. The T1 PRI channel division allows one 64 Kbps signalling channel to be used for 23 B-channels (or even, if special methods are supported, more). This allocates approximately 2783 bps per B-channel for signalling. In the case of E1 PRI channel division, there is one channel for signalling in support of 30 B-channels (giving approximately 2133 bps per B-channel for signalling).

Obviously, from these numbers, 8 Kbps is much more than is needed for normal signalling needs. BRI thus allows use of the SAPIs within LAPD addressing, to further break the D-channel into more logical channels. This allows support of packet services over the D-channel without degrading signalling support.

At the data link layer, error-free transmission is supported by sequencing and retransmission. However, the design of ITU-T Recommendation Q.921 (LAPD) was specifically created for use of separate logical channels for different purposes. Thus, we have a DLCI (combined TEI/SAPI) that is used for management interactions between user equipment and the network. We have the ability to conduct parameter negotiations and to set up TEIs and other data within LAPD. It also has the flexibility for continued expansion to allow for support of switched frame relay or cell relay.

For our purposes, however, only two of the DLCI groupings come into play. These are the signalling logical link and the broadcast signalling logical link. These are routed to the Q.931 protocol module for further examination.

CONFIGURATION OF THE SYSTEM PARAMETERS

Most ISDN user equipment has to be configured before use. This is necessary to give the equipment sufficient information to handle protocol elements on a real-time basis. (If the user had to be prompted for every piece of necessary data while a call was in progress, it is likely that the protocol would time-out because human responses are much slower than those expected, and normal, in digital connections.)

What kinds of information are necessary and why? The type of TEI is necessary (automatic or fixed). Most user equipment will use an automatic TEI, but users of routers may often employ a fixed TEI. A fixed TEI decreases the initialization time before a connection can be processed. However, it will also limit the equipment (without reconfiguration) to a specific point-to-point interface with the system.

Another piece of data that is often required is the directory number (or numbers) associated with the line (often, rather than always, because some ISDN specification variants allow autonomous retrieval of parameters from the network—after all, the network already knows most of the information that you are going to enter into the equipment.) This is used for two purposes. It is used (especially in multipoint configurations) as a check to make sure that an incoming call is meant for the equipment. It may also be supplied as the calling number for use in Caller ID applications.

As mentioned before, the purpose of the SPID has evolved in North America such that most networks now require it if data services are to be used. The SPID is often directly related to the directory number, but is not the same. The TEI, dialing number, and SPID are all directly needed for signalling purposes. The final piece of data often needed (especially for the United States) is the particular switch variant being used.

Other data desired at the administration of the device will be in the category of bearer service support. One common input is the protocol chosen. This may affect the actual SETUP message sent out to initialize the call—or it may just tell the equipment what protocol software to use after the call has been established. This will be discussed at greater length in Chapter 9.

Depending on the bearer service chosen, there may be need to provide additional information, including node addressing information, passwords, encryption code methods, compression algorithms, addressing ranges, and so forth. Generally, this information will be based on the bearer service requested.

Q.931 MODULE ENTRY AND EXIT POINTS

The software module that supports Q.931 has three logical entry points and three logical exit points. Often, these connections into the software module are implemented via operating system intertask message queues. There will be an entry from layer 2 to layer 3. In this direction, all DL_ indication and confirmation primitives will arrive. Another entry point will exist from the higher layers to layer 3. This entry point handles the N_ request and response primitives. The final entry point will often be implemented as a synchronous function (or subroutine) call. It is used for MN_ (Management Network) requests and responses.

For each of these entry points, there is a corresponding exit point. Layer 3 to layer 2 primitives (DL_ requests and responses) are sent to the LAPD module. Primitives (N_ indications and confirmations) for the higher layers are sent directly to the higher layers or, for some more complicated architectures, routed via the coordinating entity (often denoted as the C-plane in the ITU-T recommendations). Finally, management indications and confirmations are handled.

Q.931 PROTOCOL STATE MACHINE IMPLEMENTATIONS

Once the Q.931 software module has been invoked through one of the entry points, it is necessary to handle the event properly. There are various methods of implementing this, and the precise method chosen is not directly relevant to this book. However, some parts to this procedure that are common to all implementations may help the user better understand how the primitives are handled.

The first necessary task is to identify the event. Most events will be cast into the form of a primitive (even if the ITU-T recommendation does not directly state the name, and form, of the primitive). The identification of the primitive allows the software to determine what (if any) parameters are needed and to determine the methods used for parsing.

In the case of higher-layer primitives, the ITU-T does not describe particular parameters. This is highly dependent on the exact application. For example, a voice-only device may need only the dialed number as a parameter for an N_CONNECT_RQ primitive. A dedicated, "intercom" line may not even need this. However, a particular data service may need a number of parameters in order to properly specify, and service, the bearer channel to be connected.

Once the event has been identified, it must be identified in order to match it to the appropriate data structures. For example, a broadcast message may be a supplementary service message, or it may be an incoming SETUP message. It may be associated with the system as a whole or a particular call connection. One method of creating greater symmetry between various types of links is to assign data structures to broadcast links in addition to call-specific links. Although this may entail added data space that is not particularly needed in this instance, it speeds up general handling of messages by the system.

The final aspect is to parse the information associated with the event. This may be needed with any event but mainly applies to receipt of DL_DATA_IN and DL_UNIT_DATA_IN primitives. In these cases, parsing will follow the switch specifications concerning the mandatory, optional, and excluded IEs to be associated with the message type.

CALL SCENARIOS

The following call scenarios will be presented in two ways. Each will have a figure describing the associated primitives, but in an abbreviated fashion. The text will follow the sequence, going into details about the contents of the primitives. Early primitives will have most parameter details listed. Later primitives, which carry the same general function, will be described only in passing.

Each scenario should be looked at as a template rather than an exact sequence. Although some primitives will be the same for all switch variants, certain details will be different. Additionally, the exact sequence will depend very much on configuration of the user equipment as well as subscription parameters with the network.

INITIALIZATION OF THE SIGNALLING LINK

The scenario described in Figure 8.1 is for initialization of the signalling link. Initialization will always entail physical line activation (for BRI S/T this will be explicitly invoked; for PRI and BRI U the request is implicit in the interface). For most user equipment it will also include automatic TEI negotiation. In the United States and some other parts of North America, it will include SPID negotiation. This scenario is a "more complete" one that assumes TEI negotiation and SPID/EID exchange.

The initiation of the sequence begins with a DL_ESTABLISH_RQ primitive being sent to layer 2. This primitive may be sent as a consequence of a need to establish a connection, or it may be sent as a side-effect of starting the system. The former is more likely in Europe; the latter is more likely in the United States. The DL_ESTABLISH_RQ primitive has two required parameters: the SAPI and the connection endpoint suffix (CES). The SAPI is needed because this primitive is used for different types of links and the SAPI identifies the access point that is to be established. The CES is an internal (not passed to the network) identifier that can be loosely mapped to the DLCI—it identifies the purpose of the link and allows the layer 2

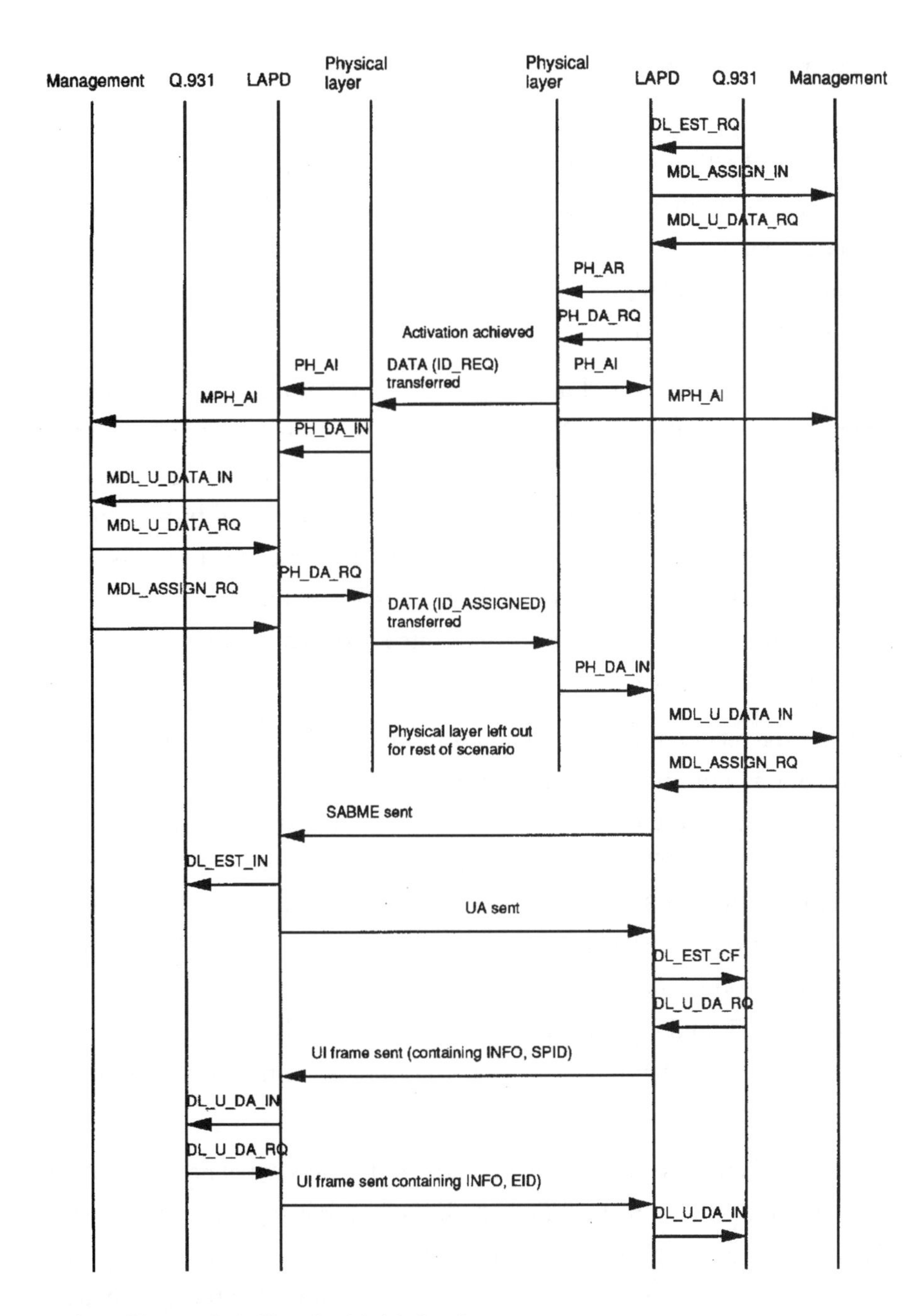

Figure 8.1 Terminal initialization.

data structures to be set up to route information after the link has been established.

The DL_ESTABLISH_RQ primitive is accepted by LAPD. It, in turn, will note that this SAPI and CES are not currently in use. It will allocate data resources for a new link. As part of this, it will set up the default state for the link. Thus, the DL_ESTABLISH_RQ primitive will be treated within the TEI_UNASSIGNED state.

Since this is a new link and no TEI has been established, the actions of LAPD are to send an MDL_ASSIGN_INDICATION primitive to the management entity and change states for the link to the ESTABLISH_WAIT_FOR_TEI state. In the case of a fixed TEI, the management entity will be able to respond immediately with a TEI value. In this example scenario, this is not possible. In the case of automatic TEI negotiation, a timer will also be started to make sure that a lost frame is regenerated. (Remember the discussion about the need to regenerate UI frames.)

The management entity receives the primitive, looks into its configuration information, and finds that this is a need for an automatic TEI. It will format a management entity identifier (MEI) message and send it to LAPD using an MDL_UNIT_DATA_RQ primitive. (MDL_ requests are sent via the TEI 127, SAPI 63 broadcast management link that is available for use in all LAPD states.) The contents of this message contain an MEI header, a two-byte pseudo-random number, a primitive identifier indicating an ID_REQUEST, and a parameter indicating what TEIs are acceptable. Generally, this last parameter must be set to indicate that any automatic TEI is acceptable.

The pseudo-random number that is part of the message is necessary because TEIs are not yet assigned. Thus, only broadcast messages may take place. This allows for the possibility of more than one terminal requesting a TEI at the same time. The pseudo-random number allows for the later matching of response to request to make sure that they correspond.

The MDL_UNIT_DATA_RQ arrives at LAPD and is processed as a broadcast management request—which is state independent. The actions of LAPD are to take the data and put an LAPD header in front of it indicating the address (TEI 127, SAPI 63) and the type of message (UI frame). It will then attempt to send the frame via a PH_DATA_RQ primitive.

The next point is a divergence that depends on the exact implementation. ITU-T Recommendation Q.921 is vague on the exact time that a PH_ACTIVATE_RQ should take place. If it is triggered by a PH_DATA_RQ primitive when the interface is currently inactive, then there must be a place for the data of the PH_DATA_RQ primitive to be queued. This may take place in either the LAPD or the low-level driver (LLD) that is handling the PH_DATA_RQ. It is also possible for the data request to be discarded until activation is achieved. The recommendations do not specify—but the various options have slightly different responses during testing conditions.

For this scenario, we will assume that the LLD will queue the data request. However, first we return to LAPD. It is ready to send a PH_DATA_RQ and it notices that activation has not yet been achieved. It sends a PH_ACTIVATE_RQ primitive to the LLD, followed by the PH_DATA_RQ primitive.

The LLD receives the PH_ACTIVATE_RQ and starts a physical layer sequence to activate the link. (The exact details are not relevant to this book.) Assuming that the sequence is successful, it will send a PH_ACTIVATE_IN primitive back to layer 2 (and MPH_ACTIVATE_IN to the management entity). In this implementation, because the PH_DATA_RQ primitive was queued awaiting activation, activation also triggers transfer of the data across the physical link to the network.

We have finally reached the network. It receives a broadcast management message asking for a TEI to be assigned. If we assume that its tables are properly reinitialized (this may not be true if the terminal had recently terminated a prior session), the response of the network will be to assign a

TEI and send back an MEI frame indicating the TEI number to be used. The pseudo-random number in this message will match that used in the request.

The LLD receives the frame and sends a PH_DATA_IN primitive to layer 2. Note that minimum information for a PH_DATA_RQ or PH_DATA_IN primitive is the address and size of the data. For a PH_DATA_RQ primitive, additional information is needed for use in "back-off" sequences needed to avoid contention on a passive bus.

LAPD receives the PH_DATA_IN primitive. It parses the address and finds that it is a broadcast management UI message. It forwards the data contents to the management entity with an MDL_UNIT_DATA_IN primitive.

The management entity checks the message and verifies that the pseudo-random number matches the request sent out. It cancels associated timers. It then formats an MDL_ASSIGN_RQ primitive and sends it to LAPD.

LAPD gets the message. This time, the message is meant for a particular link—the one for which a DL_ESTABLISH_RQ was received earlier. Thus, the primitive will be handled according to the ESTABLISH_WAIT_FOR_TEI state. Within this state, the effect of an MDL_ASSIGN_RQ primitive will be to format a SABME command frame, start a timer, and change to the AWAIT_ESTABLISHMENT state (note that we have skipped over state 4, the TEI_ASSIGNED state).

LAPD sends a PH_DATA_RQ primitive. It arrives at the network, which has already set its data link layer into the TEI_ASSIGNED state (it did the assignment). Within this state, the arrival of a SABME command calls for a UA (unnumbered acknowledgment) response frame. This is sent by the network to the user equipment via a PH_DATA_RQ, and the network's data link layer enters into the MULTIPLE_FRAME_ESTABLISHED state (state 7).

The LLD receives the data and forwards it to LAPD as a PH_DATA_IN. LAPD parses the message and treats the UA frame as the proper response within the AWAIT_ESTABLISHMENT state. The appropriate actions to be taken include canceling the original response timer and starting an "idle" timer. It will change state for the link to the MULTIPLE_FRAME_ESTABLISHED state and send a DL_ESTABLISH_CF to the network layer. Note that the CES in the initial DL_ESTABLISH_RQ established the routing for this message.

The Q.931 module now becomes involved. (It may have been peripherally involved earlier if an N_CONN_RQ triggered the initial DL_ESTABLISH_RQ.) It receives the DL_ESTABLISH_RQ and examines its state. In our scenario, we assume that this particular user equipment has a SPID and needs to get an EID.

Thus, the response of the Q.931 software module is to send out a non-specific (CRV length 0) signalling message (TEI as assigned, SAPI 0). For most switch variants, the packet is of the INFO message type with an IE indicating the terminals's SPID. This packet is formatted and a DL_DATA_RQ primitive is sent to layer 2.

LAPD receives the DL_DATA_RQ primitive for the link (now in MULTIPLE_FRAME_ESTABLISHED state), formats an information (I) frame, and sends it to the network via a PH_DATA_RQ.

The network receives the Q.931 INFO message, checks for a valid SPID, and allocates an appropriate USID and TID. This is sent back using another INFO message with an EID IE.

The LLD forwards the data to LAPD, which, in turn, forwards the data to layer 3, Q.931. (Note that only the data contents are ever forwarded; the layer 2 header is stripped as part of the function of the data link layer.) It receives the EID, makes sure that it is legal, and stores it for future use. If an N_CONN_RQ was stored while establishment was taking place, it will now be acted upon.

Initialization is now complete. The physical layer is activated. The data link layer has the signalling link established (TEI assigned and MULTIPLE_FRAME_ESTABLISHED state reached). The network signalling layer has an established indication from the data link layer and also a successful SPID/EID exchange. The system is now ready for active use. One final note: In the case of variants that require two signalling TEIs, this entire scenario is actually duplicated. However, in such an architecture, the second signalling link can be distinguished by a separate CES value.

SUCCESSFUL CALL SETUP

In the remaining scenarios, we will primarily deal only with the network layer, as reflected in the figures, which have the physical layer removed. Another aspect, however, will be increased. This will be the actions of the peer level (network layer on the other side). Thus, reciprocal actions are indicated for most of the scenarios. An outgoing call on one side is looked at as an incoming call on the other side. Similarly, this reciprocity will hold for releasing a call as well as refusal. This approach will allow us to reduce the number of scenarios explained as well as to give a more complete understanding of the full actions involved.

We will also make a number of assumptions. One assumption is that the physical layer remains active. Another is that the physical link is sufficiently secure so that data link layer errors will not occur. A final assumption is that no protocol-related errors are produced either by the user equipment or the network. (The user equipment must allow for errors, or deviations from the official specifications, from the network.)

We will not go into detail about the various potential error scenarios. These are numerous, but mainly they come into play only in a testing/certifying environment, not within day-to-day operation. The one scenario concerning refusal of a call may be encountered as well as a problem in connection parameters so that the switch refuses the setup.

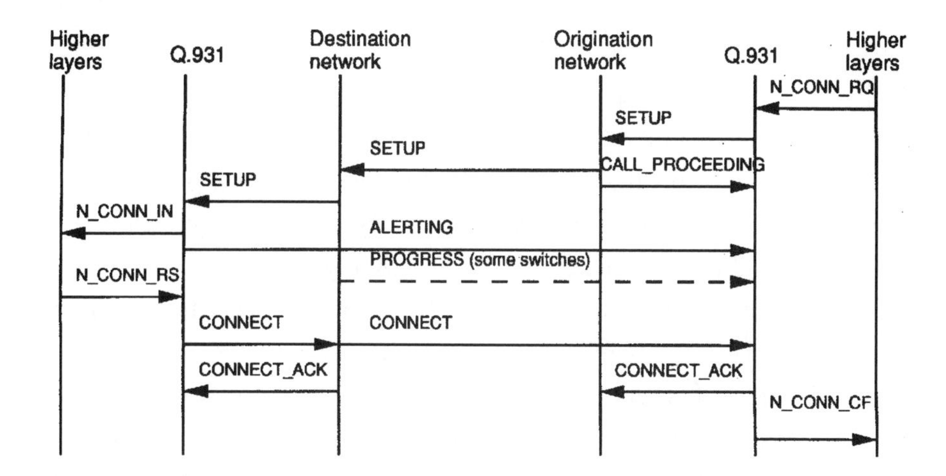

Figure 8.2 Example call set-up and acceptance.

Figure 8.2 shows the scenario for a connection request. It begins with a command from the higher layers. This N_CONNECT_RQ primitive is transmitted to the layer 3, Q.931, software module. The information in the message will contain, at least, a dialed number in this scenario. We will also assume that it specifies a form of data service.

The N_CONNECT_RQ is received by Q.931. It examines the request, notes the primitive type, and examines its own internal state. Since the signalling link has already been established, it is ready to process the primitive. It allocates a new data structure for use by the link and also allocates a CRV. (The originator is always responsible for CRV allocation.)

The original state of a link is always the NULL state (state 0). In response to an N_CONNECT_RQ, the actions will be to format a SETUP message type packet and send it to the peer. It will then make a transition to the CALL_INIT state (state 1).

The contents of the SETUP message will be the Q.931 protocol header (value 8), followed by the CRV length (1 for this scenario for a BRI link) and the CRV just allocated. The message type will be a SETUP. It should be mentioned at this point that IEs, within a codeset, always are to be presented in ascending order according to the numerical value of the IE identifier.

The first IE thus required in the SETUP message will be the bearer capability IE. This IE informs the switch as to what type of services will be required. Three of the primary bearer service types used are speech, 3.1 kHz audio, and unrestricted data. The first two types are used for speech or modem use; the third is used for digital data protocols. It is also possible to specify the data protocol more precisely in this IE. This will be covered in greater detail in Chapter 9. The bearer capability IE is a variable length IE. Therefore, the length will follow the IE identifier. In the case of unrestricted, clear-channel, 64 Kbps data, this will consist of two bytes. In the case of speech, three bytes are necessary. The third byte specifies the type of digital encoding being used on the bearer channel. These methods are specified more fully in ITU-T Recommendation Q.711.

The next IE needed is the channel identification IE. This IE may either give a specific channel (this may, in some switch variants, include the D-channel for packet use) or just say "whatever is desired by the network." Following this is the dialed number, which may either be placed into a called number (CDN) IE or a keypad IE. Generally, the keypad IE is used within North America, and the CDN IE is used elsewhere.

Other IEs may be inserted as part of the SETUP message depending on information given by the N_CONN_RQ primitive, the configuration of the system, and the specific bearer service used. The network layer now sends off the SETUP within a DL_DATA_RQ associated with the signalling link.

The network examines the SETUP message. If all necessary information is present and acceptable, it will send a CALL PROCEEDING message back to the originator of the call. It will also forward a SETUP message, as a broadcast signalling message to the peer.

Note that, due to interworking needs, it is possible that some parts of the SETUP message may be modified by the network before forwarding. In particular, within the United States, if a 56 K connection is needed before reaching the destination, the bearer capability IE may be modified to indicate this fact to the destination. The originator may be informed of this situation by use of a progress indication IE within the CALL PROCEEDING message.

The originator receives the CALL PROCEEDING message, checks to see that it is a legally formatted message, and changes state to OUTGOING_CALL_PROCEEDING state (state 3). In this state, it is able to receive continued progress, completion, or release messages but does not expect to receive any further initialization messages.

The receiving side gets a DL_UNIT_DATA_IN primitive from its data link layer. Note the asymmetry here in use of the DL_DATA_RQ primitive on the originating side and the use of the DL_UNIT_DATA_IN on the receiving side. This occurs because the originator always has a TEI at the time of origination, but the receiver may not have yet obtained a TEI. The arrival of the incoming SETUP message will signal it to obtain a TEI. The destination network, of course, knows whether a TEI has been obtained. However, the origination network does not know this. The original I-frame sent by the originator has been accepted and acknowledged. A UI-frame was generated to pass on to the destination end. The destination network could, if it wanted, change this to an I-frame before giving it to the destination terminal, but this is extra work that should be resolvable by the user equipment at the receiving end.

The receiving end gets the SETUP message. We will assume that it has already obtained a TEI (and SPID/EID exchange, if required). It notes that this is an incoming SETUP message, does some preliminary parsing, and then sets up data structures for the incoming link. The CRV is stored (but not allocated) with the upper bit set to indicate that the other end originated the call. (It will be used in this way for the rest of the call, keeping within the value the allocation responsibility.)

It now checks to make sure that a series of conditions are met. First, all mandatory IEs must be present. Second, the call must be meant for it (dialed number appropriate or matching EID). Third, it must be able to support the bearer service indicated. Finally (or first, depending on how one looks at it), it must have the general resources necessary to support the call.

If all of these conditions are true, then the ITU-T recommendations allow two separate options. One informs the higher layers of the incoming call but does not acknowledge the call with the network. The other sends back an immediate ALERT message and informs the higher layer via an N_CONNECT_IN primitive. The ALERT, as mentioned previously, is roughly equivalent to an analog ringback to indicate that the user has been informed of an incoming call. The first case has a transition to CALL_PRESENT state (state 6). The second case has a direct transition to CALL_RECEIVED state (state 7). We will assume that we have gone directly to CALL_RECEIVED state and have sent back the ALERT message.

The network now receives the ALERT message. Most switches, at this point, will send back a PROGRESS message to the originator of the call. This informs the originator of ringback—that the destination is being informed of the incoming call. (It may also inform of interworking situations.) Optionally, it will also (or, in some cases, instead of) send the ALERT on through back to the originator. If it does send an ALERT message on back to the originator, the originator will change state to CALL_DELIVERED state (state 4). Both situations can happen, depending on the particular implementation and switch variant.

We are now in a stable state. Assuming that the ALERT message was sent on through, the originator is in CALL_DELIVERED state (state 4) and the destination is in CALL_RECEIVED state (state 7). No changes are expected without higher layer input. One exception to this is that the network (and, optionally, the user equipment) can start a timer to wait for some decision on the call. After this timer expires, the call is terminated. This keeps the network resources from being tied up for an indefinite time. Since the network will use this timer, it is redundant in the user equipment and is often not implemented.

In this scenario, the destination decides to accept the call. This is done by the higher layers sending an N_CONNECT_RS primitive to the network signalling protocol. The Q.931 software receives this primitive and formats a CONNECT message. This is sent to the network, and a change to CONNECT_REQUEST state (state 8) occurs.

The network responds to the CONNECT message by sending back a CONNECT_ACKNOWLEDGE message. This takes the destination to the ACTIVE state (state 10). The network also forwards the CONNECT message to the originator, which responds with a CONNECT_ACKNOWLEDGE message (for most switch variants), and makes a transition to the ACTIVE state.

We are now ready to use the bearer channel. A note should be made here as to exactly when the bearer channel is physically available. One IE, the channel identification IE, is used to respond to the user equipment regarding which channel has been allocated for bearer services. Once this IE has been sent to the user equipment, the bearer channel is available to and from the network node. However, it may not be available all the way to the destination. This end-to-end connection does not happen until the destination accepts the call (and sends a CONNECT). Thus, the bearer channel is available at a time that is not quite in sync between the two endpoints (because the originating equipment does not know exactly when the destination network has received the CONNECT message).

Generally speaking, this becomes irrelevant due to the very small times involved. It may create some lost frames at the beginning of bearer channel use or a delay for resynchronization. Most of the time, the "best" time to allow use of the bearer channel is when the CONNECT is received. This will be slightly early for the destination equipment and slightly late for the originating equipment, but it provides for the shortest out-of-sync time for data use.

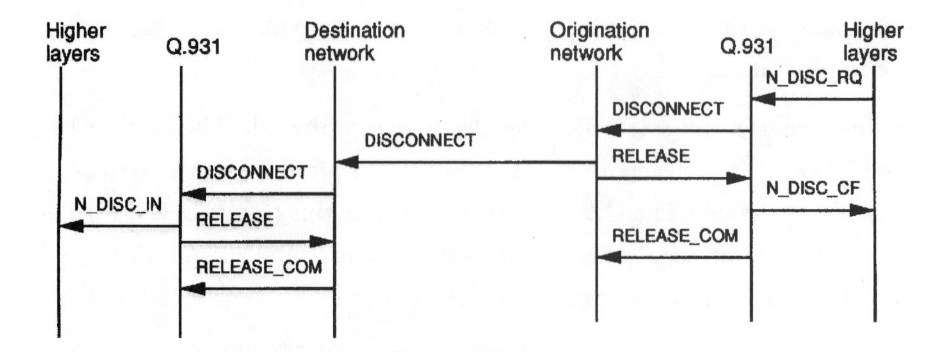

Figure 8.3 Example call teardown.

SIMPLE CALL TEARDOWN

The next scenario, as shown in Figure 8.3, is that of a simple disconnection, or teardown, of a call. We start the scenario with the call up and active. Each peer network module is in the ACTIVE state, the data link layer is still in MULTIPLE_FRAME_ESTABLISHED state, and the physical layer is still activated.

The disconnection starts from one of the higher layers (except for the form of the CRV, it really doesn't matter which end starts the disconnect, so call the side that starts this Side A and the other endpoint Side B). Side A sends an N_DISCONNECT_RQ primitive to the Q.931 module. This module, in turn, examines the primitive and formats a DISCONNECT message type. The main required IE with this message is a cause IE, which indicates the reason for the disconnection. It will then change to the DISCONNECT_REQUEST state (state 11) and start a timer to safeguard against not receiving a reply. Note that, in France, an N_RELEASE_RQ primitive may be used instead. Except for a difference in the exact state and timer used, the general effect is the same.

The DISCONNECT is a "soft" method of releasing the line. It allows both sides to clean up the protocol in a predetermined manner. The DISCONNECT is received by the network. The network sends back a RELEASE message to Side A and forwards the DISCONNECT to the other end (Side B).

Upon receipt of the DISCONNECT message, Side B will respond with a RELEASE message and transfer to the RELEASE_REQUEST state (state

19). The network receives the RELEASE and sends back the RELEASE_COMPLETE message. This causes Side B to cancel its timer and return to the NULL state. Note that, if Side B originally started the connection, it will also have to deallocate the CRV.

Side A started this sequence. In response to the DISCONNECT message, the network sent a RELEASE message. This RELEASE message causes Side A to stop its timer and return to the NULL state. Once again, if Side A originally started the connection, the CRV will be deallocated. Both sides will need to reinitialize the data structures previously used for the connection.

During testing and certification, one extensive series of tests is concerned with putting in the correct cause IE along with the DISCONNECT or RELEASE message. The result of an incorrect cause IE is to release the call even more quickly, with an altered cause IE saying that the reason is due to an incorrect mandatory IE. The purpose of such tests is to be sure that the user equipment gives the statistical information to the network that will allow proper classification of situations.

This sequence happens so quickly, with normally no human intervention after the initial request to terminate, that the point in time that the bearer channel is no longer available becomes a moot issue. Generally, the cleanest place to break the bearer channel connection is when the data structures are reinitialized. Note, however, that if a particular protocol is being used on the B-channel, the "soft" method of closing the link would be to tear down the in-band protocol before doing the out-of-band disconnect. This is similar to what is offered from MacPPP for a "soft" or "hard" close. A hard close just shuts off the link; a soft close informs the other end that the link is about to be closed.

REFUSAL OF A CALL SETUP

There are two possible simple scenarios for an uncompleted call setup. In the one instance, the call is simply refused by the other side. The network is satisfied with the request, and it meets all of the requirements by the

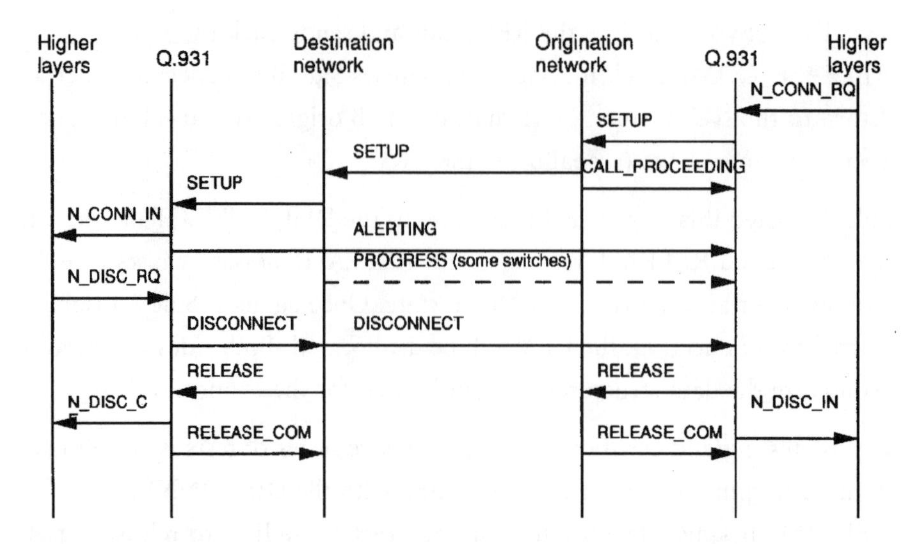

Figure 8.4 Refusal of a call set-up.

other equipment, but the call is not accepted. The second scenario involves an unintentional refusal by the network or by the other end. Figure 8.4 shows a simple refusal by the other end.

In this scenario, which starts out very similar to that of call setup, messages are exchanged that get the receiving side to the CALL_RECEIVED state and the originator to the CALL_DELIVERED state. This time, however, the terminating (or it could be the originating side if they were quick enough) side decides to refuse the call. It does this by sending an N_DISCONNECT_RQ (or N_RELEASE_RQ for some variants) to the Q.931 module.

From here on, the scenario assumes the form of a regular disconnect. The N_DISCONNECT_RQ primitive is received by the Q.931 software module. It formats a DISCONNECT message with attendant IE (or IEs), starts its timer, and goes to the DISCONNECT_REQUEST state. The network responds to the DISCONNECT with a RELEASE message, and the DISCONNECT is also forwarded to the other side. That side responds with a RELEASE message and the network sends a RELEASE_COMPLETE. The refusing side also sends back a RELEASE_COMPLETE to the network, and the call has been terminated before establishment has ever occurred.

UNINTENTIONAL UNSUCCESSFUL CALL SETUP

Unintentional refusal of a call setup is much more abrupt. It will normally involve the sending of a RELEASE message or even a RELEASE_COMPLETE message. No figure will accompany this scenario because it is a very simple set of messages in either case.

If the SETUP message is refused by the network, the network will normally respond with a RELEASE_COMPLETE. A RELEASE_COMPLETE indicates that the connection no longer exists. Thus, when the RELEASE_COMPLETE is received (and, in this case, the cause IE may be very important to determine the reason for the refusal), an N_DISCONNECT_IN primitive is sent back to the originator along with the reason. The originator received an N_CONNECT_RQ, sent a SETUP message, received a RELEASE_COMPLETE message, sent back an N_DISCONNECT_IN primitive, and returned to the uninitialized link state. No message needs to be sent back to the network—and, indeed, it would not work because the CRV is no longer valid and all data structures associated with the call have been reinitialized.

If the other end is doing the refusal, then *it* will send a RELEASE_COMPLETE back to the network, which will forward it to the originator, who will respond in the same way as above. In both instances, this automatic refusal occurs on the first message received by the network or by the other end. A DISCONNECT is normally used when the link is active. A RELEASE is used (except in some switch variants) when the release is involuntary but after the connection has progressed past the initial messages. A RELEASE_COMPLETE occurs either as an initial response (or final, in exchange with a RELEASE message) or when some catastrophic event has occurred that requires immediate release of the link. This will normally mean either a network or user equipment failure or violation of the protocol.

SUMMARY

The use of Q.931 signals for call management consists of a set of messages. Each is preceded by a protocol identification header and some form of identifier for the call to be addressed (which may be a null reference in case of system-oriented messages). Each includes a number of information elements that provide mandatory, and optional, information to be associated with the message.

Message types important in call establishment include ALERTING, CALL_PROCEEDING, PROGRESS, CONNECT, CONNECT_ACKNOWLEDGE, and SETUP. A final message type, not discussed in the scenarios, is that of SETUP_ACKNOWLEDGE. This is used by the network when a valid, but incomplete, SETUP message has been received.

For disconnection, the DISCONNECT, RELEASE, and RELEASE_COMPLETE messages are used. DISCONNECT is used as an indication that the connection is no longer needed. RELEASE is used as a request for the connection to be torn down; RELEASE_COMPLETE is an indication that the link is no longer valid.

The next chapter will go into greater detail about bearer services—which form the primary purpose behind the signalling mechanisms.

CHAPTER

9

ISDN BEARER SERVICES

Bearer services are the purpose of ISDN. Yes, it's good to be able to place a call in less than a second. It's also useful to be able to attach various data to the primitives, so that extra information (such as credit card numbers or calling card numbers) may be included directly into the connection request. These are conveniences that both help equipment users and decrease the overhead for network providers.

However, the main purpose of ISDN is to gain access to the two channels (for BRI) of 64 Kbps (uncompressed) data. In the ITU-T structure, three types of services are available: teleservices, bearer services, and supplementary services. These are fully defined in ITU-T Recommendations I.112 and I.140. Basically, bearer services relate to the bottom three layers of the OSI model.

Bearer services work as carriers of data. The teleservices combine the upper layers with the bearer services to provide access to features. Telephony is a teleservice, and speech is a bearer service. The teleservice adds flexibility and features to the bearer service—such as being able to call a specific number. Supplementary services are just that, supplements to existing service. Basic telephony allows a user to place a call. Supplementary services allow him or her to place a call on hold and reuse the same circuit (or B-channel, in the case of digital services) for another call while the first is on hold.

Supplementary services often incorporate the concept of "call appearances." A call appearance is a virtual connection. Data structures are still allocated and, possibly, a part of a connection is still in existence, but the full end-to-end connection may not be present. Call appearances must, by necessity, be mapped onto real bearer channels before they are of active use. Thus, the first call appearance may be given to B-channel 1, and the second call appearance may be allocated to B-channel 2. However, if a third call is desired (outgoing or incoming), something must be done to free up a B-channel for allocation. One of the current calls may be placed on hold, or it may be transferred or just disconnected. At this point, the B-channel may be used for the third call appearance.

Call appearances allow for identification of connections past those associated with physical lines. This identification exists for analog phones also but, most frequently, only for up to two appearances on a physical connection. It is difficult on an analog line to signal identification of more than two appearances. If one hears an interrupt tone that indicates another call is coming in, the person can signal the network (usually by a single flash of the on-hook) to change calls. The first is automatically put on hold, and the physical line is allocated to the second. Each flash allows transfer back and forth—or a conferencing of both where the same physical line is allowed coexisting use.

Most supplementary services are applicable only to voice services, although these may be used in conjunction with data. For example, a two B-channel data transfer is taking place. Supplementary services may allow an additional incoming notification and, if accepted, cause the data rate to drop down to one B-channel while the other call is accepted. This additional call is very likely to be voice.

BEARER SERVICES

The ITU-T breaks bearer services into two categories. Eight services currently are defined for circuit-mode connections, and three services are defined for packet-mode. Three of the eight circuit-mode bearer services are considered to be "essential"—every ISDN switch should be able to support them. Table 9.1 lists these 11 services. They are briefly discussed in ITU-T Recommendation I.230. Circuit-mode services are detailed in I.231.1 through I.231.8, with packet-mode services described in I.232.1 through I.232.3.

Table 9.1 Circuit-mode and packet-mode bearer service categories, adapted from ITU-T Recommendation I.230.

ITU-T Reference	Description
	Circuit-mode bearer service categories
I.231.1	64 Kbps unrestricted, 8 kHz structured
I.231.2	64 Kbps, 8 kHz structured, usable for speech information transfer
I.231.3	64 Kbps, 8 kHz structured, usable for 3.1-kHz audio information transfer
I.231.4	Alternate speech, 64 Kbps unrestricted, 8 kHz structured
I.231.5	2 × 64 Kbps unrestricted, 8 kHz structured
I.231.6	384 Kbps unrestricted, 8 kHz structured
I.231.7	1536 Kbps unrestricted, 8 kHz structured
I.231.8	1090 Kbps unrestricted, 8 kHz structured
	Packet-mode bearer service categories
I.232.1	Virtual call and permanent virtual circuit
I.232.2	Connectionless
I.232.3	User signalling

The recommendations mentioned in I.231.6 through I.231.8 are oriented primarily toward PRI. These are known as H0, H11, and H12 channels and are used as aggregate B-channels. As we will see in examination of the bearer capability IE in the next section, Table 9.1 does not cover all possible permutations.

You will note that all of the circuit-mode services include the requirement for "8 kHz structured." This is an indication that a separate timing signal accompanies the transmission that ensures that the data will always arrive in byte form. This is particularly important if any type of hardware bonding is to occur with the B-channels.

ITU-T bearer service types I.231.4 and I.231.5 *are* primarily oriented toward BRI, but they are not considered to be "essential." That is, ISDN nodes are not required to support them. As is often the case, something not required is not supported. This even must be the case in a situation where the network operates in a multivendor environment. An optional service becomes unusable unless both endpoints, and all network nodes in between, support the service.

The primary purpose of these two services is multiuse capabilities. The first allows speech and data to exist on the same bearer channel. Rather than having to use supplementary services to place one connection on hold, voice and digital data are each marked, in some manner, directly within the data stream. Speech signals can be routed to a transceiver, and data signals can be routed to a protocol stack. Because more advanced digital encoding methods (such as those described in ITU-T Recommendation G.721) allow speech encoding in a 32 Kbps signal, both could actually share the line simultaneously. It would also be possible to use the line in a half-duplex mode (similar to the way many speaker phones work). The potentials can be seen but the equipment needed would be much more specialized; so far, networks are not implementing the service.

The other non-essential service would be of particular interest to users of ISDN with the Internet. This allows both B-channels to be connected with a single SETUP message. Currently, there must be two separate SETUP messages involved, each connecting a particular B-channel. Although this method certainly works, it seems likely that this non-essential service will be offered at some point in time. Note that this does *not* imply that bonding will be done by the network. There will still need to be some form of hardware or software (such as ML-PPP or X.25 MLP) bonding in use to combine the two B-channels.

Let's now examine the use of the three "essential" bearer service types. These are usually referred to as unrestricted data, speech, and 3.1 kHz bearer services. Unrestricted data means that the network will not alter the data as it passes through the network (with the caveat that, with 56 K interworking, the data may be required to be in a certain form).

The speech service type allows the greatest network flexibility. It allows the network to use various network mechanisms such as echo cancellation, intermediate analog transmission, and other "normal" analog voice network handling of the data. The data are not guaranteed to be exactly the same as when transmitted, but the general quality of the signal is required to be sufficient for non-degradation. In other words, this service applies the same quality standards as for analog voice circuits (and will often be of higher quality except for situations where much of the circuit is going through analog networks).

The 3.1 kHz is favored for use when analog equipment is going to be used over the bearer channel. It is much more restrictive on the network concerning how the data may be manipulated. This means that the data arriving should be essentially the same as that which is transmitted. The permissible level of degradation allowed for speech is not allowed within this service.

BEARER SERVICE NEGOTIATION

There are two ways of negotiating the bearer service to be provided. One is out-of-band, using the IEs defined within Q.931. The other is in-band, taking place after the connection is set up. This in-band negotiation may take place over any type of bearer service, but it is most likely for services that are oriented toward data services—unrestricted data and 3.1 kHz bearer services.

The original intent of the ITU-T (then CCITT) was to have services negotiated using the IEs of Q.931. This allows the greatest structured use of services. One side indicates the type of service that it wants to use. The other side decides to accept, or it negotiates with the other end for something that both sides can handle.

Much existing ISDN equipment has been designed using the in-band approach, probably due to several factors. The Q.931 method allows tremendous flexibility of options. These options enrich the protocol set but may cause problems for equipment that does not implement exactly the same set of options. As we will see shortly in the next subsection, the Q.931 method allows specifying the protocol down to very small details. If all manufacturers do not agree on the details that will be supported (and the exact format of these details), then interoperability becomes more difficult.

Another reason for in-band negotiation is that the standards are always created slowly and changed slowly. In-band negotiation allows for new features to be used quickly. A final reason lies more in the area of competition—if a manufacturer offers an appealing set of services using its own in-band negotiated methods, then it will become popular and force other manufacturers to use similar methods and so become a de facto standard.

Q.931 BEARER CAPABILITY USE

Out-of-band service negotiation for ISDN is based on the bearer capability IE. This IE, as seen in Figure 9.1, is a variable-length IE. The first four bytes are considered to be mandatory and must be included in all bearer capability IEs—which, in turn, must be in any SETUP message. The first byte is the identifier for the IE, and the second byte is the length. The third

Bits

8	7	6	5	4	3	2	1	Bytes
0	0	0	0	0	1	0	0	1
	Bearer capab ility information element identifier							
Length of the bearer capability contents								2
1 ext	Coding standard		Information transfer capability					3
1 ext	Transfer mode		Information transfer rate					4
1 ext	Rate multiplier							4.1 (multirate)
0/1 ext	0 Layer 1 ident.	1	User information layer 1 protocol					5*
0/1 ext	Sync/ asynch	Negot.	User rate					5a*
0/1 ext	Intermediate rate		NIC on Tx	NIC on Rx	Flow control on Tx	Flow control on Rx	0 Spare	5b* (V.110)
0/1 ext	Hdr/ no hdr	Multi frame	Mode	LLI negot.	Assignor /ee	In-band neg.	0 Spare	5b* (V.120)
0/1 ext	Number of stop bits		Number of data bits		Parity			5c*
1 ext	Duplex mode	Modem type						5d*
1 ext	1 Layer 2 ident.	0	User information layer 2 protocol					6*
1 ext	1 Layer 3 ident.	1	User information layer 3 protocol					7*

ITU-T Recommendation Q.931 has many additional notes

* - optional

Figure 9.1 The bearer capability IE, adapted from ITU-T Recommendation Q.931.

byte specifies the information transfer capability that incorporates the essential service types and some other ones not directly mentioned in ITU-T I.230. This third byte also contains the coding standard—one of which is the ITU-T standardized coding, which shall be used in this discussion.

The information transfer capability includes values for speech, unrestricted digital information, restricted digital information, 3.1 kHz audio, unrestricted digital information with tones/announcements (formerly called 7 kHz audio), and video. With only six capabilities currently defined, there are 26 additional codes that may be used in the future.

The fourth byte is concerned with the transfer mode and rate. Currently, only circuit-mode and packet-mode are defined for the transfer mode (allowing two more values to be defined in the future). One thing that we see, and will continue to see in these discussions, is that the recommendations always allow for options not yet agreed upon. The transfer rate is concerned with the number of 64 Kbps channels to be set up for this call. One option is that of "multirate," which allows a variable number of B-channels to be set up in a given connection. This is used in conjunction with the first optional byte specifying the number of B-channels to be used.

At this point, we need to discuss just how optional bytes are used within an IE. A byte can only be optional if there is some way to understand it. That is, if bytes 4.1 through 7 are optional, how do we know that we are now looking at byte 6 with the preceding optional bytes omitted or at byte 4.1? This is done in two ways. One is by context. The value indicating "multirate" is an example of this. This value means that the next byte (4.1) is mandatory. A second method is by use of a fixed code used in a field. We see this method in bytes 5, 6, and 7 in the identification field. This can be used to distinguish between the three possible fields. The third is through use of the extension bit in the high-order position. A 0 indicates that this field is not ended. Thus, within the fields, bytes may be omitted but not skipped. If byte 5 is used (indicated by the 01 in the identification part) and

5c is needed, then bytes 5a and 5b must also be included (but 5d can be omitted if not needed).

The only additional bytes relevant in this discussion are bytes 5 and 5a. Byte 5 is concerned with the physical layer and is needed to determine the basic formatting of the signal on the bearer service. Thus, the digital encoding method of speech service must be coded here if speech is the bearer service. In the case of unrestricted data, this *may* be used to indicate the type of protocol being used (such as V.110, V.120, and X.31, which will be discussed later). Byte 5a is useful in interworking situations because it can indicate a lower data rate than that of the transfer rate. For example, the transfer rate is 64 Kbps, but the user rate is only 56 Kbps if the network has had to interwork with a 56 K network. This rate information can be put in by the network if necessary. Some of the ramifications of this will be discussed later.

The use of the bearer capability IE allows only declaration of use; it does not allow for actual negotiation. Negotiation is done by use of two other IEs called the low layer compatibility (LLC) IE and the high layer compatibility (HLC) IE. In Europe, the LLC (if present) must be a mirror image of the bearer compatibility IE within the SETUP message. In the United States, the LLC IE is often not passed through the network (thus, precluding negotiation and giving another reason why so many manufacturers use in-band negotiation techniques).

The LLC IE (and, in Japan, the HLC IE) is most important in later messages. The LLC IE may be optionally used in two ways for negotiation. First, it may be responded to with a matching, or differing, LLC IE contained within the destination endpoint's CONNECT message. For example, the originating terminal says that it wants to use V.120. The destination does not support V.120 and wants to use V.110 instead. It indicates this in an LLC IE contained within the CONNECT message. The originator can then accept, or decline, service with this new LLC.

A second method is by use of multiple LLC IEs contained within the SETUP message. According to Annex J of ITU-T Recommendation Q.931, the originator may include up to four LLC IEs (normally presented in order of preference). Thus, the originator may present a set of alternative capabilities to the destination, and the destination indicates the agreed-upon set of attributes with the LLC IE contained in the CONNECT message.

IN-BAND BEARER SERVICE NEGOTIATION

Three forms of in-band service negotiation are possible. The first is that of prior agreement. This is generally the same as what basic analog service users do at present. If you want to send a fax, you dial a fax number. If you want to speak with someone, you dial a voice number. A modem connection requires a modem to be connected to the other end. There is no actual negotiation because both ends use a specific protocol.

The second, which makes use of analog transfer techniques, is used when analog equipment is used over a speech or 3.1 kHz bearer service. One algorithm in common use is a sequence of a time delay, carrier detect, and final service. In other words, if the receiver is picked up during a specific period of ringing, the call is taken over by the voice equipment used. Next, the network checks for a carrier signal. Finally, if neither of the previous conditions has occurred, the transmission is routed to a default piece of equipment. This algorithm is useful for a phone, fax, and answering machine all operating on the same phone line.

The third method is true negotiation. Most equipment making use of true negotiation employs its own proprietary algorithm. However, we can still look at what's needed and devise a potential algorithm ourselves.

The first branch of options (on an unrestricted data bearer service) addresses whether the data protocol is HDLC-based or direct frame structure. As we will see shortly, V.110 is one such direct frame method—setting each bit in the 64 Kbps data transmission according to its own protocol needs. Another is digital voice. A third might be video. In the cases of

V.110 and most video methods, there will be specific framing patterns for which to look. For V.110, this framing pattern consists of an all 0 byte followed by nine bytes with a 1 in the low-order bit. Video will have its own framing pattern.

If there is no specific framing pattern to check (which is true for speech), then we must next check for HDLC patterns. If the HDLC format is not being followed (interframe fill with flags or markers), then the equipment can go to its default non-HDLC mode (similar to the answering machine up above).

If an HDLC protocol is detected, the next question is which one. The address field is the first field following a frame delimiter. If it has only one byte (the extension bit set to 1 in the first byte) then it is *not* V.120 or LAPF. It may be X.25 layer 2 (LAPB). If it has two bytes, the address subfields may be checked to see what values are present. Certain values are allowed for different protocols (with the defaults normally not occupying overlapping values).

Checking for patterns is how in-band recognition is accomplished. The negotiation requires both ends to be able to perform the same in-band algorithm. The originator always gets to give the first preference. However, it must continue to check for either a valid response (to its preferred data protocol) or an alternative proposed by the other equipment.

INTERWORKING

When users in the United States think of interworking (a short form of internetworking) they usually think of a situation where the 64 Kbps channel capacity is reduced by a 56 K connection somewhere in the transmission path. However, interworking takes place any time that two different networks, with potentially different physical interfaces and signalling protocols, work together.

One common international interworking situation is for speech or 3.1 kHz bearer service use. In both of these situations, it is necessary to designate the digital encoding method within byte 5 of the bearer capability IE. In North America and Japan, the normal encoding method is called the μ-law. Elsewhere, the A-law is used. Both of these methods are defined in ITU-T Recommendation G.711. (A third, for speech over 32 bps circuits, is defined in G.721.) By international agreement, networks supporting the μ-law must do the conversion necessary to and from A-law using networks. This conversion also includes modification of byte 5 of the bearer capability IE, in the case of ISDN.

In any case where data have the potential for modification, it is necessary to be able to inform both ends of the change. This allows the equipment to adjust for the difference or disconnect the call. Sometimes, as in the above speech interworking case, no special actions are needed because the network has already taken care of all modifications. The destination will receive exactly what it expected, and the origination may continue to send (and receive) for what it was originally configured.

NOTIFICATION OF INTERWORKING

If a data transfer is modified in transit, both ends need to be aware of this to adapt to the situation. Within ISDN, two fields, or messages, are used for this purpose.

In the direction of the call (incoming SETUP message), the bearer capability IE is normally modified. The network may place V.110 into byte 5 and set a user rate of 56 Kbps in byte 5a. This is done by convention because V.110 (as will be seen shortly) was designed with 56 Kbps interworking specifically in mind. This does *not* preclude the use of any other data protocols. However, unless the LLC IE is passed unchanged, we will have a situation that requires prior agreement of protocol use, since the out-of-band negotiation has been lost during the overlay of the bearer capability IE by the network.

The originator of the call must also be notified. This can be done in two different places: as part of a progress indicator IE in a call establishment message that is returned by the network (CALL PROCEEDING, SETUP ACKNOWLEDGMENT, ALERTING, or CONNECT) and in a progress indicator IE in a separate PROGRESS message that is sent by the network (after ALERTING, if used). This progress indicator IE contains a "progress description" field that permits various interworking notifications. Code value 1 is of specific importance as it indicates "not end-to-end ISDN." Code values 2 and 3 are also of interest as they indicate that one end is not ISDN.

Note that the one end may be a 56 K endpoint or it may be an analog device. A 56K endpoint may be dealt with in the same manner as interconnecting changes in the middle of the network. Analog equipment should not be connected (the call should be refused by the network) if an unrestricted data bearer service was requested. (Note that speech may be degraded in this situation if a 56 K interworking situation occurred and the endpoint is an analog speech device.)

METHODS OF 64 KBPS TO 56 KBPS INTERWORKING

The data capacity of a circuit can only be as great as the lowest speed available on the line. Thus, in the case of 56 Kbps interworking, data throughput has been throttled back to 56 Kbps. This reduction of capacity must be allowed for at both ends.

Both methods of reducing data rates incorporate the idea that only seven of the eight bits of each byte may be used for valid data. By using this method, the data throughput has been reduced by 1/8—to the desired 56 Kbps transfer rate. Normally, the high-order bit is the one unused by the endpoints. This occurs because the 56 K network will "bit-rob" this location for signalling methods (that is, this bit is appropriated for other uses than for end-to-end data transfer).

One method is to use a data transfer protocol that already allows for the bit to be overwritten. V.110 is one such protocol, which is why the network may make use of the protocol within byte 5 of the bearer capability IE as part of interworking notification. We will discuss the format and use of V.110 later in this chapter.

The other primary method for adaptation to this interworking situation is to make use of hardware support to send and receive data employing only reliable bit positions. Motorola, Siemens, and other semiconductor manufacturers allow for this in the command structure of some of their devices. This works by specifying just which bits of each byte received and transmitted will carry valid data. Data are then placed into the transmitting frame, bit by bit, into the appropriate bit fields.

As an example, say that we have the pattern

```
01111110111001101100111001111110
```

to be transmitted using only seven bits of each byte. Assume that the above string of bits is listed in transmission order (physical line protocols usually transmit low-order bits first, so each byte is "inverted" according to how it is used by software at each end). We now have the following pattern being transmitted (once again, in transmission order):

```
0111111X0111001X1011001X1100111X1110011X...
```

Note the additional 011 that has been tacked onto the end of the original data. This is the beginning of a new interframe fill character. Data are always continuous, even though user data may not be in progress. Thus, the data have been transmitted on a 64 Kbps circuit but using only 7/8 of the available local capacity.

Any method of adaptation to interworking conditions must be agreed upon by both endpoints. The above two methods are ones that are common in current networks.

USE OF SPEECH OR 3.1 KHZ FOR DATA

Digital speech, or 3.1 kHz, bearer service provides approximately the same quality and data rates available on a standard analog line. The actual data rate is much higher, but since it is an encoded version of the analog signal, the actual sampling rate works out to a theoretical 4 kHz equivalent.

Pulse code modulation (PCM) is the method used for digital encoding of the audio for most standard lines. As mentioned above, ITU-T Recommendation G.711 gives two different standards, called the A-law and the μ-law. A coder-decoder (codec) is used to convert between the PCM encoding and analog form. This is implemented with a combined chip set or integrated into another semiconductor device.

The codec allows external attachment of analog devices (phones, faxes, modems) to be used on a bearer channel. Sometimes a DTMF signal generator and detector are used in combination with the codec to allow the use of tone dialing and feedback. In this manner, the analog equipment can be used in the exact same way that it has normally been used. The off-hook, on-hook (and flashes of brief on-hook period), and DTMF signals are recognized and translated into ISDN signalling commands. Signalling in this manner is not required, but it does give full backward-compatibility with old analog equipment. Another hardware component that is sometimes included is a modem-decoder (modec), which allows internal modem use with an ISDN (or other digital) device.

UNSPECIFIED 64 KBPS CLEAR CHANNEL USE

This form of bearer service allows the greatest access to the speed of the bearer channel. As discussed above, both endpoints will need to use the same protocol for transmission and reception of the data. This may be decided by out-of-band negotiation, in-band negotiation, or prior agreement. Data use falls into two general categories—packetized and continuous stream data. HDLC protocols are an example of packetized data. Video would be a good example of continuous stream data. In the first case, the valid data are delimited by flags. In the second case, it is assumed that valid data are on the transmission line at all times.

TYPES OF DATA ACCESS

Q.931 provides the signalling mechanisms for ISDN. The bearer channels give the physical conduits for data transfers. One final component is needed to actually use ISDN for data purposes—that is to provide access to the bearer channel. In other words, there must be some method to transfer the data from the host (or other data-producing entity) to the bearer channel.

The data themselves may be handled asynchronously (or one byte at a time) or synchronously (in groups of data). This distinction affects the choice of an interface, but there are still many different possible access points. A traditional method is through a serial port—just as modems are currently used. This means use of the ISDN equipment as a terminal adapter (TA) at the R interface reference point, discussed earlier in this book.

Another method for internal devices is that of using shared memory or data bus access. In this manner, the data are transferred as if they were located in local memory from the host side and the ISDN equipment side. This allows data transfer rates limited only to the bus access times—which usually means that the limitations are irrelevant in comparison to those of the physical interface. For PRI equipment, the systems may need to be tuned for full data transfer.

A final category is that of special ports, such as the codec output port (which usually look like a modular phone jack). Other special devices may be used in conjunction with standards such as the multi-vendor integration protocol (MVIP), which allows data distribution between equipment cards.

SHARED MEMORY OR BUS ACCESS

Shared memory access is very simple—but not necessarily easily portable. An area of RAM on a device is memory mapped into both the host and ISDN device. (The amount of protocol software on the ISDN device may be as much as the first six OSI levels or as little as the physical interface chips.) The ISDN device makes connection with the data bus of the host to allow full memory access by both devices. Hardware logic is then used so that accesses within a particular range of addresses are responded to by the memory contents of the RAM. The RAM is equally accessible by the ISDN device or just by the hardware doing the I/O (perhaps via a direct memory access, or DMA, between the hardware device and the shared memory).

The lack of portability is concerned with the absence of standards for this application. While the memory may be shared, and contention (both trying to read or write at the same time) controlled by memory access logic, a protocol is still needed to let each side know when the data are ready. Many different techniques are possible.

If the protocol stack is located on a separate processor, the shared memory may be used for primitives. The primitives can be based on ITU-T primitives. A "mailbox" is often used for such a configuration. One side puts data into the memory area reserved for primitives. When that side is ready, it puts the flag up on the mailbox. It is not allowed to put any additional primitives into the memory area until it has been acknowledged (the flag is put down again). Note that this may be a positive (primitive accepted) or negative (primitive denied) acknowledgment. This approach works as long as both sides agree to the protocol. The "flag" acts as an interprocessor semaphore.

Another possibility, for "dumb" cards that have specialized semiconductor devices on them but no general-purpose processor is to memory map the registers of the devices. This allows control of the semiconductor devices directly from the host. Data can either be transferred serially or via shared memory accesses.

SERIAL PORT ACCESS

Serial port access is a common method of data transfer in today's communications systems. External modems are attached to the serial ports of computers. Most PC communication protocols assume either an external modem via a serial port or an internal modem accessed by communication port (COM for DOS machines) driver protocols.

Serial port access is limited in speed. It also requires an intelligent device on the other side of the R interface reference point because commands (often in the de facto AT command set) must be translated into ITU-T primitives or directly into Q.931 signalling messages. Another requirement may be to packetize data if a packetized protocol such as X.25 or V.120 is used.

Note the difference between serial port access and analog port access. The serial port is the same as that used to connect a modem to a computer. The analog port is the same as the connection to the analog network via a wall jack. Serial port access does not imply the form of the data format. Analog port use does indicate the form of the data being used across the interface.

OTHER PORT ACCESS

ISDN devices are not limited to just the access methods that have been used in the past. In fact, due to the much greater bandwidth available (particularly with PRI or even B-ISDN), it may be expected that new methods and ports will be developed. We have already discussed analog port use to a certain extent.

The MVIP is potentially very useful in ISDN because it allows the integration of interface devices with data processing devices. The current standard allows for an 8 x 2 Mbps extended data bus between device cards. This may also be broken down into 256 64 Kbps data streams—in other words, about the capacity of 10 T1 lines. This is still not quite up to B-ISDN standards (the lowest category of which, so far, is a little more than 25 Mbps), but its versatility is tremendous.

SPECIFIED BEARER SERVICE USE

Even if the bearer capability IE is not used for protocol specification, an examination of some of the possible values will give us a better idea of what types of services were anticipated by the ITU-T. In our discussion of byte 3, we covered the basic service types that have already been presented. We also briefly mentioned unrestricted data information with tones/announcements. This uses a modification of G.721 that allows a 7 kHz audio signal to be transported across the 64 Kbps bearer channel. Video is usually protocol specific.

Byte 5 is more important as to exactly *how* the data is to be transmitted and received. Three values are concerned with speech/audio. Three other values are associated with terminal adaption. These three values will be discussed in much greater detail.

Terminal adaption is useful to fulfill one of the primary objectives of ISDN architecture—support of existing networks and equipment. V.110 and V.120 are two different approaches to supporting "start-stop" types of terminals. When used in asynchronous mode, these protocols offer support of serial port connections in an emulation of modem use (with attendant AT command interfaces—although the command interface is not mandated as part of the ITU-T recommendations).

The third terminal adaption protocol listed as a value for byte 5 is that of X.31. X.31 allows use of ISDN with existing X.25 networks. Unlike the byte-stream data applications for which V.110 and V.120 were designed, X.25 is a packet-based protocol.

X.31

ITU-T Recommendation X.31 is not a data protocol. It is an integration protocol used for employing of X.25 within ISDN. X.31 covers are two distinct situations. The first, called Case A, occurs when ISDN is used to set up a conduit between the user and an X.25 router. The second situation, called Case B, exists when ISDN handles X.25 packet services itself.

These two scenarios are common within the ISDN recommendations that are oriented toward keeping the flexibility of location of service provision. In other words, it is good for ISDN to be able to offer various data services. However, it is also necessary that ISDN be able to allow the user access to services that it does not provide itself. Case A and Case B are used in frame relay SVC provision also.

Cases A and B

Case A has two parts to it for user equipment. The first part is to make the connection to the X.25 router. This is done by use of a SETUP message that specifies only unrestricted data access (it is not inherently limited to 64 Kbps for a connection). The scenario for call setup, given in the previous chapter, applies to this connection.

The second part exists after connection to the X.25 router has been made. At this point, the ISDN is no longer directly involved. It does not examine the contents of the bearer channel—allowing transparent access by the user and X.25 router. The X.25 protocols are now required. (It may actually be ITU-T Recommendation X.75, which is an extension of X.25 oriented toward international use.) The protocols within X.25 will be discussed shortly. For now, it is sufficient to know that a second connection will be made.

In Case B, the X.25 packet handler is part of the ISDN. Because this is the case, the normal IEs associated with the SETUP message are involved if it is an on-demand service. On-demand service indicates that access to the service has been subscribed to, but that it is not available full-time.

Most network support of X.25 on the D-channel is on a "nailed-up" basis—it is available on a semipermanent basis whenever the terminal has an active connection to the network. However, since the channel identification IE is available within the SETUP message, it is also possible to have the D-channel as part of the connection parameters for an on-demand call. Both semipermanent and on-demand X.25 connections may fall into Case A or Case B situations. A semipermanent connection for Case B, non-network-provided data services, is equivalent to a leased line except that only the bearer channel or channels are available.

X.25 Basic Protocol

ITU-T Recommendation X.25 is an HDLC-based protocol. As such, it is frame delimited, and the data link layer comprises an address, control, and (if necessary) data field. X.25 is an older protocol than ITU-T Recommendation Q.921; thus, in many ways Q.921 (LAPD) is based on what was discovered about the needs of the network through many years of X.25 network use.

X.25 has only one address byte. This is sufficient because additional logical links are provided at the network layer. The C/R bit is present, but the bit uses are fixed based on the role being played by the equipment. X.25 was designed as an equipment-to-network protocol. It divided these roles into data terminal equipment (DTE) and data communication equipment (DCE) roles. The DTE uses a fixed value to indicate commands and another value for responses. The DCE uses the opposite values for similar purposes. This dichotomy is required.

This dichotomy also causes a lot of problems when the network is not directly involved—when one terminal is communicating directly with another terminal. This situation is common within ISDN. It is partially addressed by X.75 as well as the ISO document 8208.

Another interesting feature of the data link layer of X.25, or LAPB, is that of the concept of the multilink procedure (MLP). Although it is not a popular protocol for end-user data usage, it is a very early method of software bonding of multiple channels. Thus, it is an historical basis for other bonding methods. The MLP consists of a header that is interposed between the data link layer header and the network layer data. As such, it is similar to V.120 header bytes, and it may be considered an extension to LAPB or part of the data handled by the network layer. Due to software considerations, it is easier to think of it as an LAPB extension.

Because LAPB and LAPD have so many similarities, they are often implemented by the same software module for an integrated ISDN TA. The extension bit in the first byte may be used as a differentiator between LAPD or LAPB. The control field is basically identical, and the state events and responses are very similar. One major difference exists: There is really no concept of a TEI within LAPB. Thus, only Q.921 states 4 through 8 are applicable to LAPB.

The network layer header, as shown in Figure 9.2, has four major fields. These consist of the general format identifier (GFI, consisting of the higher nibble of the first byte, the group and channel ID, packet identifier, and additional data, if needed.

The GFI allows indication of modulo use. X.25 is defined for both modulo 8 and modulo 128. Only the modulo 8 form is given in Figure 9.2. The top two bits are used, for data packets, as additional qualifiers for the data. In fact, the Q bit is designated the Qualified-data bit. The D-bit allows for user-to-user acknowledgment of data transfers rather than just network-layer-to-network-layer.

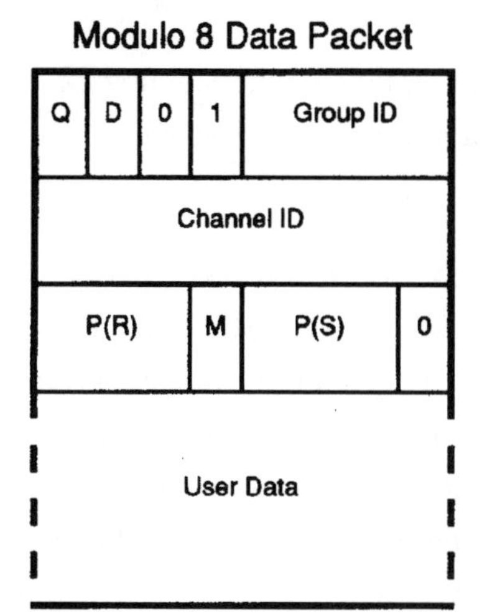

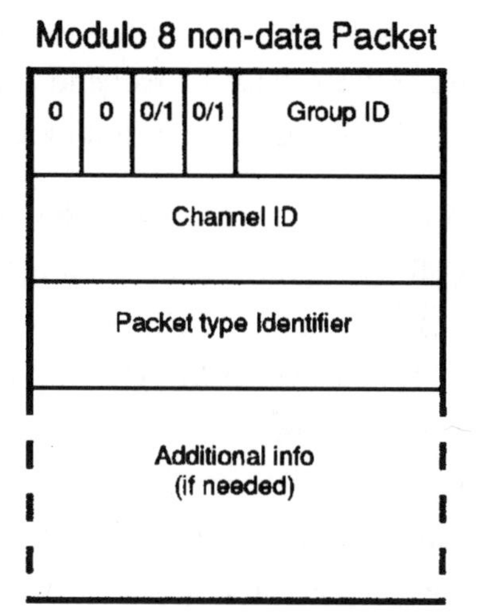

Figure 9.2 X.25 network layer packet formats, adapted from ITU-T Recommendation X.25.

The channel identifier is used in three ways. It can identify the in-band signalling channel. It can be allocated for a new logical link being supported. It can be pre-assigned as the identifier for a permanent virtual circuit (PVC). The algorithm for assignment of new logical link addresses deals with ranges of logical channel identifiers (LCIs).

The group identifier is a method by which terminals may be considered as a subnetwork. They may be able to only talk with another. They may have certain features allowed. One group may be able to receive incoming calls only. This may also be called a closed user group (CUG).

The packet type identifier uses the low-order bit to distinguish data packets from non-data packets. If the identifier is set to 0, it is a data packet and the remaining bits give sequencing information, just as in LAPD. The M-bit can be used for segmenting larger data blocks into packets acceptable by the X.25 network and reassembling them before passing them to the receiver's higher layers.

The X.25 packets fall into four categories. These are call setup and clearing, restart, flow control and reset, and data and interrupt. The first two will be discussed next, followed by the data and interrupt packets.

X.25 Signalling Protocols

Call setup and clearing messages include the CALL REQUEST, CALL CONNECTED/ACCEPTED, CLEAR INDICATION/REQUEST, and CLEAR CONFIRMATION. As may be inferred from the naming conventions, the same messages are used for outgoing and incoming messages. An incoming packet indicating that a CALL REQUEST has been connected is called a CALL CONNECTED message, while the CALL ACCEPTED is a message sent in response to an incoming CALL REQUEST. This change of name based on the direction of the message can become confusing until one remembers that the actual message is the same.

The protocol is very simple: send a CALL REQUEST and wait for a CALL CONNECTED. Note that, just as in Q.931, if the data link has not been established, the network layer will need to send a DL_ESTABLISH_RQ primitive to LAPB (or LAPD in the case of a D-channel packet handler). A CLEAR REQUEST is acknowledged with a CLEAR CONFIRMATION (or DCE clear confirmation, to give its full name). Timers are used, but the effect of the protocol is to treat call setup and clearing messages as the equivalent of UI frames at the data link layer.

The RESTART and RESTART CONFIRMATION messages are normally used by the terminal in two instances—when a Case A X.31 link has just been set up and when only PVCs are allowed on the channel. ISO 8208 also gives a procedure by which negotiation of the DTE/DCE role may take place between two pieces of terminal equipment. The network may also initiate a RESTART.

X.25 Packet Layer Protocols

Two types of data packets exist within X.25. One is just called DATA, and the other is called an INTERRUPT packet. Data packets operate in the same manner as do I-frames at the data link layer. An INTERRUPT packet acts like a combination of an I-frame and a UI-frame. Like a UI-frame, it has priority over any I-frames that are still waiting to be acknowledged. It is also unaffected by any windowing mechanisms currently in use. However, like an I-frame, it has an acknowledgment—the INTERRUPT CONFIRMATION message. Because these two messages are linked one-to-one, it also indicates that only one INTERRUPT can be allowed at a time. Sometimes this message is considered to be an "expedited data" frame—a data frame with a high priority.

The flow control and reset category of messages consists of RR, RNR, REJ, and RESET messages. The RR, RNR, and REJ messages operate similarly to those of the data link layer. The RESET message has an effect similar to a SABME at the data link layer. The use of RESET may cause data loss that the higher layers may need to accommodate.

Note that, for data transmission, there is a considerable redundancy in sequencing and allowing for errors. In fact, layer 3 protocols assume that the data link layer will not recover. This redundancy is what causes a much greater overhead compared to newer protocols—and it is why frame relay technology has become popular. X.25 was created in a period of very unreliable networks and physical transmission lines, and many of its protocol precautions are no longer needed today.

V.110

V.110 and V.120 are both specifically designed for use with ISDN. The framing rate for V.110 was designed around the 64 Kbps bearer channel capacity. V.120 has a subset of Q.931 messages available for in-band call setup. One item of interest is that V.110 was designed with the capacity for direct interworking with 56 Kbps networks; however, it is primarily used in Europe where 56 Kbps interworking is largely unnecessary.

ITU-T Recommendation V.110 falls into the category of continuous data use of the bearer channel. This does not mean that valid data are on the channel at all times. Rather, the 64 Kbps channel is under the control of the protocol and there are no interframe periods to be filled. Table 9.2 shows the framing format for a 2400 bps 80-bit intermediate frame. Note that the bits are shown in transmission order (inverted from the normal software order) with the lowest order bit on the left. Some data (D) bits are repeated. There are also E bits and S bits. The E bits are used in synchronous modes. The S bits are used to report the status of V.24 interface circuit leads.

V.110 is negotiated via the bearer compatibility IE or via prior agreement. It is important that each endpoint be aware of the exact speed used by each side (although the speeds do not have to be symmetric, there is no specific method of negotiating a receiving and a transmitting speed). V.110 is capable of synchronous, or asynchronous, data transfers, but asynchronous use is much more common.

Table 9.2 V.110 2400 bps intermediate frame format, adapted from ITU-T Recommendation V.110.

0	0	0	0	0	0	0	0
1	D1	D1	D2	D2	D3	D3	S1
1	D4	D4	D5	D5	D6	D6	X
1	D7	D7	D8	D8	D9	D9	S3
1	D10	D10	D11	D11	D12	D12	S4
1	1	1	0	E4	E5	E6	E7
1	D13	D13	D14	D14	D15	D15	S6
1	D16	D16	D17	D17	D18	D18	X
1	D19	D19	D20	D20	D21	D21	S8
1	D22	D22	D23	D23	D24	D24	S9

Because V.110 operates by imposing a specific framing standard onto the bearer channel, it is often supported by a hardware device. Once the channel has been set up, the receive and transmit data leads are connected directly to the device, which then fulfills the requirements of the protocol. It can be implemented in software, but the real-time demands are considerable and force an upgrade in processor power needed. Thus, the use of software-supported V.110 is not always the least expensive alternative.

One of the primary tasks for any terminal adaption protocol is that of rate adaption. Rate adaption is the carrying of a specific data stream (which goes across the R interface reference point) to that of the data channel carrier. V.110 does this in three steps. The first step, referred to as Rate Adaption 0 (RA0), applies only to asynchronous use. This allows the data to be sufficient to fill in an 80-bit intermediate frame.

Note, from Figure 9.3, that the frame has 24 actual data bits within it. The intermediate rates are 8000, 16000, and 32000 (64 Kbps is not considered to be an intermediate rate because it is the same speed as the channel). One hundred 80-bit frames may be transmitted in a second for an 8000 bps intermediate rate speed. By using only 24 out of the 80 bits, the rate is adapted down to a 2400 bps user data rate. This adaption step is called RA1.

The bearer channel, however, is 64 Kbps. How is the 8000 bps adapted to the 64 Kbps channel? It uses the same method mentioned for general interworking with 56 Kbps networks—it does not use all of the bits of each byte. For an intermediate rate of 8000 bps, only the low-order bit of each byte is used. For an intermediate rate of 32000 bps, only the bottom 4 bits are used. This partial use of the bytes is called the final adaption step (RA2). It is immediately obvious why V.110 is so easily adaptable to 56 Kbps interworking.

V.120

V.120 is an HDLC method of terminal adaption. As such, it may be used with a packet assembler-disassembler (PAD) software module as part of a

TA function within ISDN equipment. (It may also be used in a shared memory situation for greater data access.) Rate adaption is fulfilled as a natural part of the HDLC protocol. A data frame is created based on the data currently available for transmission. The channel contains interframe filling until more data are ready for transmission. Thus, no special adjustments need to be made for rate adaption.

V.120 is defined as a "delta" document from that of Q.921. As such, it only marks differences from LAPD. Generally, the data link layer is used in the same way as LAPD with a few important differences. One is that the SAPI and TEI are used as a combined logical link identifier (LLI) rather than as separate fields (however, it is a one-on-one overlay onto the two LAPD fields). A second is that the command and response bits are fixed. A command (from either side) is marked with a C/R value of 0 and a response is indicated by a 1. Realizing that some implementations of the data link layer may be restricted in their use of the C/R bit, V.120 also allows I-frame responses in addition to commands.

V.120 has two specific advantages over other terminal adaption protocols. One is that it allows for in-band signalling using a subset of the Q.931 messages and states. The other advantage is that, in out-of-band setup or in-band, it is possible to specify different protocols to be used at layers 2 and 3. (However, layer 2 must be Q.921—with V.120 extensions, if in-band call control is to be used.)

In addition to directly allowing for rate adaption, HDLC protocols have another advantage over direct framing patterns. This advantage is that the data rates do not have to be constant—nor do they have to be the same in both directions. A caveat to this statement is that the physical connection across the R interface reference point *will* be set to a particular bit speed. Thus, differences are permissible in the actual data sent across the bearer channel, but the average effect will still be the user data rate specified at the

R interface reference point. This relies on TA buffering and the ability to use something special within V.120 called control header bytes.

One control header byte is mandatory with V.120. This is called the Header (or H-byte) byte. It carries across terminal data from peer to peer. This data includes "break" data, error indications, and segmentation data used for synchronous data transfers. A second header byte is optional. This byte is called the control state information byte and contains mirror-images of RS-232 control leads. This facilitates use of flow control across serial interface ports.

The in-band signalling component of V.120 is rarely used. It supports only states 0, 1, 7, 10, and 19 from the user set of protocol states. It uses message types that are very familiar in comparison to those used within X.25 call control. It uses only the SETUP, CONNECT, RELEASE, and RELEASE COMPLETE message types (parallel to CONNECT REQUEST, CONNECT ACCEPT, CLEAR INDICATION, and CLEAR CONFIRMATION of X.25). It uses these message in the Q.931 standard manner, with a couple of additional IEs supported to allow for LLI allocation and protocol negotiation.

INTERNET-SPECIFIC BEARER SERVICE USE

X.31 is not really directly relevant to an Internet user on ISDN. It is, however, extremely important as an example of other types of data services that may be desired on ISDN equipment that the user will need to purchase. Just as most modems are used to connect to a variety of services, ISDN equipment should allow for the variety of services the user may need.

In particular, users of ISDN equipment will probably want voice service (although they may retain their analog equipment and use an analog port on the ISDN equipment). (They can also retain a separate analog line.)

They may want to take advantage of Group 4 facsimile transmission, which allows fax transfers to take place at speeds of more than six pages per minute. Note that Group 4 faxes are often done in combination with X.25 (or X.75 in international situations).

However, this book is about Internet use over ISDN. In this direct area, the terminal adaption protocols (V.110, V.120, and some others not discussed in this book) are of use. General unrestricted data use is also directly relevant.

The terminal adaption protocols are used as methods to connect to Internet nodes using character stream-oriented protocols. This is very similar to the access used by people with modems and with a character-mode account. It is also possible to do synchronous protocols such as point-to-point protocol (PPP) if an asynchronous-to-synchronous software module is available for conversion.

PPP is a successor to the SLIP protocol. Its original intent was to allow multiple protocols to be transported in a router environment. PPP, in this manner, is an enhancement to SLIP and allows ISDN connectivity in addition to older network protocols such as ethernet.

PPP is carried over the bearer channel as an HDLC protocol. The Q.931 SETUP message uses the unrestricted 64 Kbps data mode. (Interworking with 56 Kbps networks is still possible using the same methods discussed earlier.) PPP is primarily an encapsulation protocol. Its purpose is to identify the protocol that is being carried and to route the data appropriately from one location to another.

The transport control protocol/Internet protocol (TCP/IP) is the primary protocol suite actually used for Internet access. It is called a suite because it encompasses a number of specific protocols—each used with particular networks and for specific services. It is normally packet-based and thus could be used directly over a system that supports HDLC framing. This requirement (and that of PPP or SLIP) means that synchronous communication is in use—which is why a converter is needed at the R interface if such is used directly.

If the R interface *is* used, then the speed bottleneck becomes that of the physical interface. In a shared-memory situation, the speed limitation is based on the bandwidth of the connection. In the case of PRI, this may mean an H0 (384 Kbps) channel or an individual B-channel. For BRI, two B-channels are available but (currently) only as two separate conduits.

In order to make use of these multiple conduits, PPP has had extensions added (within request for comment, RFC, documents) to make use of multiple channels. It does this through the methods of windowing and sequencing that we have discussed about LAPD (and referred to about MLP within X.25). This allows maximum capacity and greater versatility in use over ISDN or other combinations of networks.

SUMMARY

Bearer services provide the transport services, and access to the speed, of ISDN. They may be negotiated out-of-band, via Q.931 information elements, or they may be negotiated in-band. The simplest form of negotiation is that of prior agreement. In the United States, partially because of interworking situations with 56 Kbps networks, equipment manufacturers often use in-band negotiation.

The architecture of ISDN is oriented toward the continuing support of existing networks. Doing this also indicates support of existing equipment. This is done by use of various interworking standards, as well as making use of the various reference points as locations where conversion, between the existing protocols and those needed by ISDN, takes place.

Terminal adaption protocols are particularly useful for accessing the Internet. However, these still must face physical limitations imposed by the support of older equipment. Use of unrestricted data bearer services allows other protocols to be imposed on the channel. These may be in support of non-Internet services or for those directly applicable to accessing the Internet.

PART

3

RESOURCES

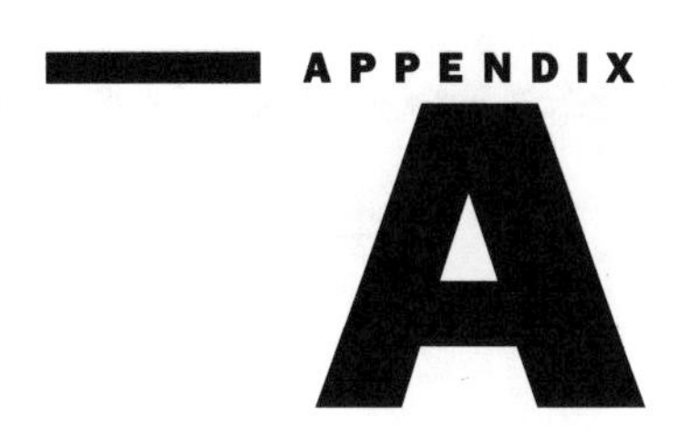

LOCATING ISDN INFORMATION ON THE INTERNET

Use your favorite search engine to locate information on ISDN. Please note that Web addresses may change from time to time; therefore, the following information is subject to change.

GENERAL INFORMATION ON ISDN

Name	URL
Dan Kegel's ISDN Page	http://alumni.caltech.edu/~dank/isdn/
Bellcore	http://www.bellcore.com/pub/isdn
USENET ISDN FAQs	comp.dcom.isdn http://cis.ohio-state.edu/hypertext/faq/usenet/isdn-faq/top.html

REGIONAL BELL OPERATING COMPANIES

Name	URL
Ameritech	http://www.ameritech.com/products/business/asg-ds-asds-isdn.html
Bell Atlantic	http://http://www.ba.com/isdn.html
BellSouth	http://www.atglab.bls.com:80/products-services/isdn.html
NYNEX	http://www.nynex.com (not active at publication)
Pacific Telesis	http://www.pacbell.com/products/sds-isdn/overview/isdn-ovr-toc.html
Southwestern Bell	http://www.sbc.com/swbell/shortsub/isdn
US West	http://www.uswest.com/isdn/index.html

ISDN LONG DISTANCE CARRIERS

Name	URL
AT&T	http://www.att.com
MCI	http://www.mci.com
Sprint	http://www.sprint.com
MFS	http://www.mfsdatanet.com

EQUIPMENT MANUFACTURERS

NT-1S

Name	URL
AT&T	http://www.att.com
Adtran	http://www.adtran.com/cpe/isdn/netterm.html
Motorola	http://www.mot.com
Northern Telecom	http://www.nortel.com
Tone Commander	http://www.halcyon.com

ISDN TERMINAL ADAPTER CARDS

Name	URL
IBM	http://www.raleigh.ibm.com/wav/wavprod.html
Digiboard	http://www.digibd.com
ISDN*tek	http://www.isdntek.com
KNX	http://www.fleet.britain.eu.net/~KNX/
Planet ISDN	http://nw.com/satusa/
US Robotics	http://www.usr.com/products.html

ISDN EXTERNAL STAND-ALONE DEVICES

TERMINAL ADAPTERS

Name	URL
Adak	http://www.icus.com
Adtran	http://adtran.com
3Com	http://www.3com.com
Motorola	http://www.motorola.com
Racal Datacom	http://www.racal.com
Zyxel	http://www.zyxel.com

ROUTERS

Name	URL
Ascend	http://www.ascend.com
Combinet	http://www.combinet.com
Cisco	http://www.cisco.com
Digiboard	http://www.digibd.com
Rockwell	http://www.rns.rockwell.com
Telebit	http://www.telebit.com

PERSONAL URLS FOR FUTURE REFERENCE

Organization	Uniform Resource Locator

ISDN PRODUCT DESCRIPTIONS AND TELEPHONE SERVICE ORDERING INFORMATION

Appendix B is intended as a reference source for general information on a representative sampling of ISDN hardware and, when provided by the manufacturer, information on how to order the ISDN telephone service from your local exchange carrier. It covers a variety of products including NT1s, terminal adapters, and a small number of routers and bridges. Contact information is furnished at the end of each product description. A summary of these products and their use with the Internet is provided in the main chapters of this book.

READER'S NOTES

We advise you to validate performance specifications with the reseller or directly with the manufacturer prior to purchase. Specific pricing on each product has been omitted due to the variance in price you may encounter from reseller to reseller.

When it comes time to order your telephone service the sales agent may question you about the equipment you will be using. In some cases the agent may have a copy of the manufacturer's recommended line configuration for the product. It is always a good idea to make sure the sales agent is familiar with your equipment; if the agent is not, you should refer him or her to the manufacturer or furnish the information contained in this appendix.

The authors make no claims regarding the specifications and performance data presented here. Proprietary data, trademarks, and copyrights are noted to the extent practicable.

The information presented here has been culled from company product literature. In some cases the material has been edited for suitability. Please note that we have attempted to present this material without preference toward any specific product. Remember that this appendix does not include every ISDN product on the market. New products are introduced frequently.

Although the primary aim of this book is to address the single-user ISDN dial-up requirements and solutions for Internet access, much of the material applies to network or LAN connections as well. For that reason, LAN products are included here. In almost all cases, the LAN solution will require a bridge or router. Terminal adapters, computer cards, and external standalone devices are the dominant hardware solution for the dial-up service.

LIST OF ISDN PRODUCTS

The following ISDN products are described in this appendix.

- B1. ISDN TERMINAL ADAPTERS
  - A. ADAK 220 and 221
  - B.1. ADTRAN ISU EXPRESS

- B.2. ADTRAN ISU 128
- C. IBM WAVERUNNER™
- D. MOTOROLA BITSURFR TA210
- E. 3COM/IMPACT

- B2. ISDN LAN CARDS
  - A. CISCO 740 SERIES ISA BUS ADAPTER CARD
  - B.1. DIGIBOARD PC IMAC
  - B.2. DIGIBOARD DATAFIRE
  - C. U.S. ROBOTICS SPORTSTER ISDN 128K™
  - D. ISDN*TEK
  - E. PLANET - ISDN BOARD
- B3. ISDN ROUTERS AND BRIDGES
  - A.1. ASCEND PIPELINE 25
  - A.2. ASCEND PIPELINE 50
  - B. CISCO 750 SERIES ETHERNET TO ISDN ROUTER
  - C. GANDALF 52421
- B4. NETWORK TERMINATION DEVICES (NT-1S)
  - A. ADTRAN
  - B. AT&T
  - C. MOTOROLA
  - D. NORTHERN TELECOM
  - E. TONE COMMANDER
  - F. IBM NETWORK TERMINATOR EXTENDED

B1. ISDN TERMINAL ADAPTERS

B1.A. ADAK 220 AND 221

ADAK Communications Corp.
5840 Enterprise Drive
Lansing, MI 48911
Tel. 5l7-882-5191 Fax 517-882-3194

The ADAK 220 and 221 are two ISDN access products permitting simple, low-cost connectivity to all ISDN-based voice and data communications. The ADAK 220 and 221 combine call management capabilities with high-speed digital communications. This unit is compatible with existing analog equipment (for example, analog telephones, data terminals, fax machines, key systems, security/alarm, etc.,) eliminating the need for additional ISDN dedicated equipment (terminal adapters, NT-1, and ISDN telephones).

In addition to data transport, this product supplies a robust feature set of advanced call management functions usually found in digital corporate communications systems. The ADAK 221 has an S/T digital out port which supports ISDN S/T data based and video adapters used for video conferencing. Both units allow access to the global X.25 network, offering cost-effective data communications. The ADAK 220 and 221 support ISDN National-1 (NI-1), National-2 (NI-2), Custom central office switch software, and Siemens. Applications of the ADAK 220 and 221 include the following:

- Advanced voice communications management
- High-speed digital data access to information services
- Point-of-sale credit card authorization and verification
- Access to the Internet
- Telecommuting from home or remote office
- Education-intensive distance learning

- Connection to ISDN-ready desktop videoconferencing (ADAK 221 only)
- Very secure alarm monitoring and cut-line detection
- Easy, cost-effective access to fast regional and national health care, financial, educational, and travel networks
- Customized protocols for selected data communications applications
- Cost-saving POTS-to-ISDN service cutover, allowing easy installation

SPECIFICATIONS

ISDN SERVICE OPTIONS—BASIC RATE INTERFACE (BRI)

General Capabilities

- Integrated NT-1 functionality
- Integrated terminal adapter functionality
- Automatic POTS-to-ISDN service cutover
- Backup power eight hours [optional]

Voice Capabilities

- Performs incoming call acceptance routing or rejection via caller ID; A standard analog caller ID box or display telephone can be connected to the analog lines
- Call hold, forward, conference, transfer, screening, and multiple call appearances
- Ability to activate central office switch capabilities (voice mail, call forwarding, call conferencing, etc.)
- Audio response and prompts for user interaction

DATA CAPABILITIES

- X.25 PAD access over the D-channel (X.3, X.28, X.29, and T3POS)
- Hayes compatible modem PAD for access to X.25 network
- Support for security/alarm applications (optional)
- Support for point-of-sale applications

ISDN Switch Compatibility.

Switch Manufacturer	Siemens	Northern Telecom	AT&T
ISDN Software			
National1 (NI-1)	xx	xx	xx
National2 (NI-2)	xx	xx	xx
Custom	(*)		

(*) Supports data and most call features

INTERFACES

ANALOG TELEPHONE PORTS

- Two standard analog telephone lines provided by two RJ-4C analog connectors
- Supports up to five ringer equivalencies on each line for analog telephone devices

SERIAL PORTS

- Two RS-232 serial ports supporting synchronous data speeds up to 128 Kbps and asynchronous speeds up to 115.2 Kbps (DB-25 ports)
- Reverse charging capability on data calls

S/T PORT

- An RJ-45 S/T port supports two 64 Kbps B-channels and one 16 Kbps D-channel in multipoint mode (ADAK 221 only)
- Battery-backed “Power One” supplied continuously at 1 watt, and “Power Two” supplied with wall power up to 8 watts

ISDN BRI U Reference Point

- One RJ-45 ISDN BRI for the U reference point for direct two wire ISDN access

Installation and Configuration

- Configuration implemented through a standard analog telephone interface and voice prompts with keypad input (DTMF) or serial device (PC or terminal)
- Windows configuration utility
- Local or remote configuration
- Local or remote software upgrade

LED Status Indicators

- Power status
- Two error status indicators
- Network activity/status indicator

Operating Environment

- Operating temperature: 0 to 50 degrees Celsius
- Storage temperature: –20 to 70 degrees Celsius
- Relative humidity: up to 95%, non-condensing

Note: prolonged temperatures above 30 degrees Celsius will have a notable impact on the lifetime of the optional backup battery.

Agency Approvals and Certifications

- FCC Class B
- UL 1459

Remote Support and Testing Capabilities

- Software can be upgraded via RS-232 or remote D-channel software download

- Support for remote configuration
- Extensive remote and local internal diagnostics, including complete testing of the ISDN line

Physical Characteristics

- Dimensions: 9.5" × 9.3" × 3.1"
- Weight: 3 lbs. (with optional battery: 4 lbs.)

Power Supply

- Integrated with a surge suppressor
- Includes a UPS with an optional internal battery supply eight hours of backup time

ISDN U interface	Standard
ISDN S/T interface	Standard (ADAK 221 only)
ISDN power	Standard
Two Sync/Async ports	Standard
D-channel X.25	Standard
Two POTS lines (with caller ID)	Standard
Hold/forward/transfer/routing/ screening/multiple calls	Standard
Voice prompts	Standard
X.3, X.25, X.29 PAD	Standard
Hayes X.25 PAD	Standard
Auto POTS-ISDN cutover	Standard
Remote configuration	Standard
B-channel X.25 and V.120	Standard
Custom protocol PADs	Optional
Backup battery (8 hours)	Optional
Cut-line alarm detect	Optional

Remote diagnostics	TBA
Least cost call routing	TBA
Auto self configuration	TBA

B1.B.1. ADTRAN ISU EXPRESS

Adtran Inc.
901 Explorer Blvd.
Huntsville, AL 35806-2807
Tel. 800-923-8726 Fax 205-971-8699

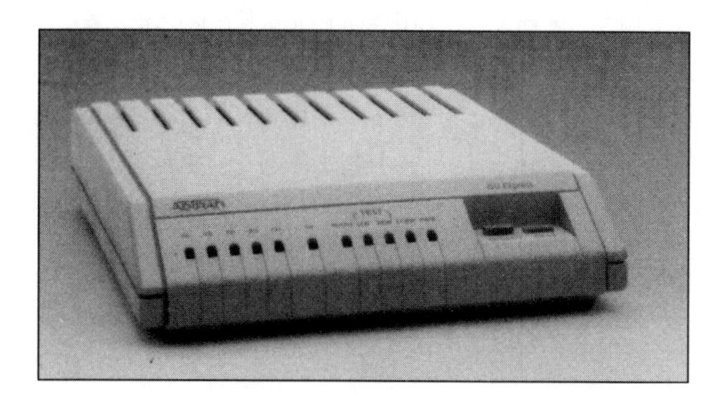

The ISU Express is a stand-alone device that connects data terminal equipment to the ISDN network. The ISU Express looks like a modem, but it uses a high-speed digital telephone line. It allows data transfer at rates up to 128 Kbps, which is much faster than a standard V.32 analog modem. This unit is a scaled-down version of the ISU 128 and is specifically designed for applications such as high-speed modem replacement, telecommuting, remote office interconnection, and Internet access.

For work-at-home or remote office applications, the ISU Express is available with an optional POTS telephone interface. This provides an RJ-l1 connection for a regular analog telephone, fax machine, or external analog modem. This feature allows a remote office to use a single ISDN line for a telephone and/or fax machine while also providing an RS-232 interface for a computer. If a call is in progress on the telephone, data can still be transferred to and from your computer at up to 64 Kbps, simultaneously. If there is no telephone call in progress, you can send data at up to 128 Kbps.

The ISU Express is also available with an optional integrated V.32bis/V.42bis modem. This option allows the ISU to communicate not only with ISDN terminal adapters or switched 56 devices but also with analog modems. In a remote office application, this eliminates the need for an external analog modem and an extra telephone line.

The ISU Express transmits data over an RS-232 interface and performs at synchronous data transfer rates from 2400 Kbps to 128 Kbps and asynchronous rates from 300 Kbps to 115.2 Kbps. At rates over 64 Kbps, the BONDING delay equalization protocol synchronizes data over the two 64 Kbps B-channels. Configuration of the unit is achieved using a VT100 menu system and AT commands. Dialing is initiated in-band over the RS-232 interface using AT commands, V.25bis, or DTR assertion. The front panel contains LED indicators for power, network readiness, DTE interface leads, ring indication/Off Hook status, and local and remote loopback testing.

FEATURES

- Supports work-at-home, telecommuting, and remote office applications
- Connect your telephone, fax, and computer to one digital dial-up phone line (ISDN)
- Dial up the Internet, bulletin boards, ISDN terminal adapters, switched 56 DSUs, and analog modems
- Supports error-free data transfers at rates up to 128 Kbps without compression
- Uses one ISDN phone line to support multiple connections such as telephone, fax, and computer
- Send and receive phone calls and transmit data simultaneously

- Built-in network termination NT-1
- ISDN line effectively replaces two regular telephone lines

ISDN Line Ordering Information

Request an ISDN Basic Rate Interface (BRI) line with U interface reference point and 2B1Q coding.

Choose one of the following:

- 2B+D service (supports up to 128 Kbps)
- 1B+D service (supports up to 64 Kbps)

The ISU Express supports the following switch types and software protocols:

- AT&T 5ESS: Custom, 5E6 and later software, National ISDN-1
- NTI DMS-100: BCS-32 and later software (PVCI), National ISDN-1 (PVC2)
- Siemens EWSD: National ISDN-1

Request that the ISDN line allocate one dynamic terminal endpoint identifier (TEI) per phone number.

For service offered from an AT&T 5ESS, request a multi-point line (2 Directory Numbers), with the following features:

Bl service:	On Demand (DMD)
B2 service:	On Demand (DMD) if 2B+D
Data line class:	Multi-point (MP)
Maximum B-channels:	2 if 2B+D, 1 if 1B+D
Circuit-switched voice bearer (CVS) channels:	Any
Number of CVS calls:	1 (recommended for testing purposes)

Circuit-switched data (CSD) bearer channels:	Any
Number of CSD calls:	2 if 2B+D, 1 if 1B+D
Terminal Type:	Type A

Turn the following features off:

- Packet mode data
- Multiline hunt
- Multiple call appearances
- Electronic key telephone sets (EKTS)
- Shared dictionary numbers
- Accept special type of number
- Intercom groups
- Network resource selector (modem pools)
- Message waiting
- Hunting
- InterLATA competition

For service offered from a Northern Telecom DMS-100, request a point-to-multipoint line, with the following features:

Line type:	Basic rate functional
Electronic key telephone sets (EKTS):	No
Call appearance handling (CACH):	No
Non-initializing terminal:	No
Circuit-switched service:	Yes
Packet-switched service:	No
TEI:	Dynamic

Bearer service: Circuit-switched voice and data permitted on any B-channel (Packet-mode data not permitted)

Select a long distance carrier and request circuit-switched 64 Kbps, clear channel access if possible.

Long distance access should be provided through your normal carrier.

Ensure that the telephone company provides the following information for configuring the ISU Express:

- ISDN switch type
- ISDN switch protocol version
- ISDN phone number(s)
- Whether the ISDN line is point-to-point or multipoint
- Service profile identification (SPID) number(s) with prefixes and suffixes, if applicable (if ISDN line is multipoint)

B1.B.2. ADTRAN ISU 128

Adtran Inc.
901 Explorer Blvd.
Huntsville, AL 35806-2807
Tel. 800-923-8726 Fax 205-971-8699

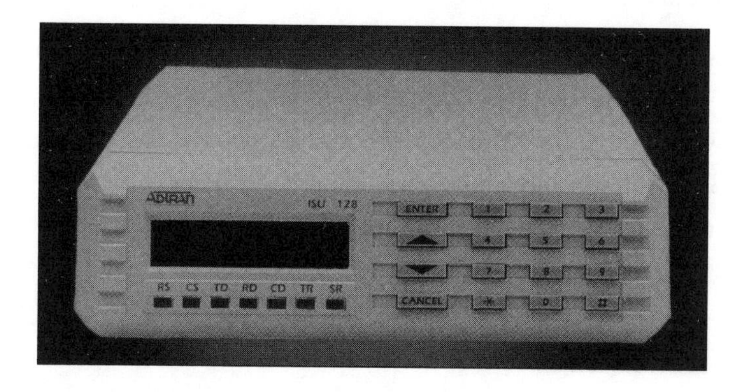

The Adtran ISDN ISU 128 is a stand-alone device that connects data terminal equipment to the ISDN network. The ISU 128 can be viewed as an ISDN modem that allows high-speed data transmission at rates up to 128 Kbps.

The ISU 128 lets you take advantage of high-speed data transmission using 2BlQ ISDN BRI service. From the network, ISDN is delivered by a single 2-wire U interface, which is connected directly to the ISU 128. ISDN network termination is designed into the ISU 128, thus eliminating the need for an NT-1. For network testing, the ISU 128 responds to loopback commands from the telephone company central office.

By using the ISU 128, the user can migrate ISDN into existing network services and data communications equipment. The ISU 128 interoperates with V.32bis modems (optional) and switched 56 DSUs as well as various ISDN terminal adapters and inverse multiplexers.

The ISU 128 transmits data over either an RS-530Aa or V.35 interface. An adapter is available for RS-232 interfaces on cable lengths less than 15 feet. The ISU 128 performs at synchronous data transfer rates of 2400 Kbps to 128 Kbps and asynchronous rates of 300 Kbps to 115.2 Kbps. At rates over 64 Kbps, the BONDING delay equalization protocol synchronizes data over the two 64 Kbps B-channels.

Dialing from the ISU 128 is accomplished by using the front panel, stored numbers, DTR assertion, AT commands, or V.25bis in-band dialing. The RS-366 parallel dial interface can be used for applications such as video-conferencing and facsimile.

The ISU 128 also supports dedicated leased 2B1Q services. This provides a dedicated point-to-point service (as in a limited distance modem or leased line application), with no dialing necessary. The ISU 128 has semi-automatic dial back-up capabilities when used in a limited distance modem or leased-line application. Upon recognition of line failure, the ISU 128 will begin connection to the remote unit over the ISDN line. The unit can manually or automatically return to the leased line when dedicated service is restored.

The front panel of the ISU 128 consists of a 2-line by 16-character display, seven LEDs, and a 16-button keypad. This allows for configuring, dialing, testing, and monitoring of the unit without data terminal or test equipment. AT commands may also be used for configuration and status monitoring.

FEATURES

- Provides up to 128 Kbps of dialed bandwidth for high-speed data applications
- Uses standard ISDN BRI service
- 2BlQ modulation conforms to ANSI Tl.601.1992 ISDN U interface standard
- Built-in network termination NT-1
- Supports dedicated 128 Kbps 2BlQ service in a limited distance modem or leased-line application
- Supports dial backup with Basic Rate ISDN when used in limited distance modem or leased-line application
- Supports synchronous data transfer from 2400 Kbps to 128 Kbps and asynchronous data transfer at 300 Kbps to 115.2 Kbps
- Manual, automatic, and stored number dialing
- Supports BONDING delay equalization protocol
- Compatible with AT&T 5ESS, Northern Telecom DMS-100, and National ISDN-1 switches
- Provides U interface lightning protection

ISDN Line Ordering Information

Request an ISDN Basic Rate Interface line, with U interface reference point and 2BlQ line coding.

Choose one of the following:

- 2B+D service (supports up to 128 Kbps)
- 1B+D service (supports up to 64 Kbps)

The ISU 128 supports the following switch types and software protocols.

AT&T 5ESS	Custom, 5E6 and later software, National ISDN-l
NTI DMS-100	BCS-32 and later software (PVC1), National ISDN-1 (PVC2)
Siemens EWSD	National ISDN-1

Request that the ISDN line allocate one Dynamic terminal endpoint identifier (TEI) per phone number.

For service offered from an AT&T 5ESS, request a point-to-point line, with the following features:

B1 service:	On Demand (DMD)
B2 service:	On Demand (DMD) if 2B+D
Data line class:	Point-to-Point (PP)
Maximum B-channels:	2 if 2B+D, 1 if 1B+D
Circuit-switched voice bearer (CVS) channels:	Any
Number of CVS calls:	1 (recommended for testing purposes)
Circuit-switched data (CSD) bearer channels:	Any

Number of CSD calls:	2 if 2B+D, 1 if 1B+D
Terminal type:	Type A

B1.C. IBM WAVERUNNER™

IBM U.S.
Dept. 2VO
1130 Westchester Ave.
White Plains, NY 10604
Tel. 800-426-2255 Fax 800-242-6329

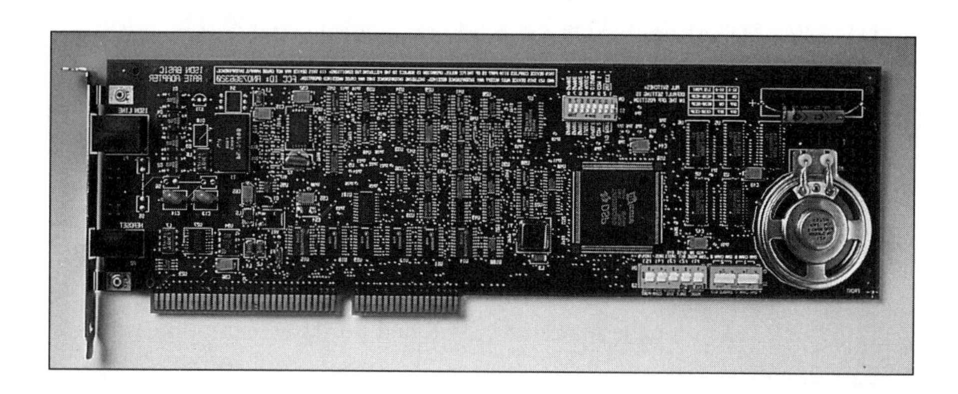

The IBM WaveRunner™ is an internal terminal adapter, full-length PC card that allows your personal computer to communicate with other ISDN terminal equipment at speeds of up to 128 Kbps. It also is capable of communicating with existing analog line equipment, such as standard modems, fax machines, and with switched 56 digital service. WaveRunner provides these capabilities on a single adapter PC card. With its analog emulation and digital interfaces WaveRunner gives the user the benefit of high-speed digital communications while preserving the interoperability with existing equipment.

WaveRunner supports a number of applications, such as Internet accesses, LAN-WAN interconnection, telecommuting and work-at-home, multimedia communications, video and data conferencing, image transmission, and high-speed file transfer. With it, you can migrate your network to ISDN in measured steps or all at once and still maintain interoperability with non-ISDN networks at both the home and office. WaveRunner is designed to accept software upgrades by using a programmable digital signal processor.

FEATURES

- Machine requirements: 386SX or higher processor, able to support DMA bus mastering, DOS 5.0/Windows 3.1 or higher (enhanced mode), OS/2® 2.1 or higher.
- Central office switches supported: AT&T 5ESS®, Northern Telecom DMS-l00™, National ISDN-1 (NI-1) compatible switches.
- ISDN requirements and specifications: ISDN Basic Rate Interface service line (user provided), NT-1 network termination equipment (user provided), passive bus, S/T interface.
- Interoperates with other ISDN equipment at speeds up to 128 Kbps, V.120, adapter at 64 Kbps, V.120, analog modems at negotiated speed (V.32bis), G3 facsimile machines, switched 56 data communications equipment.
- Standards supported:
  - Telephony: ISDN, Switched 56
  - Modems: Bell 103, Bell 212, V.22bis, V.32, V.32bis
  - Modem commands: Hayes AT
  - Data compression: MNP5/4
  - Facsimile: Group lll.
- Applications supported:
  - TCP/IP: RFC 1294, SLIP, IP packet

- Windows: Crosstalk®, DynaComm®, HCL-eXceed/W™, HyperACCESS®, PC Anywhere™, ProCOMM PLUS®, Qmodem™, Smartcomm™, TurboCom, Windows Terminal

Ordering ISDN for WaveRunner™

WaveRunner is designed for use in the United States only.

5ESS Parameters

Channel Configuration:	2B, V/C (alternate circuit switched voice and data)
Terminal Type:	A (recommended but not required)
Display:	No
Autohold:	No
Onetouch:	No
Idle Call Appearances	

DMS100 Parameters

Channel Configuration:	2B/V/C (alternate circuit switched voice and data)
Signaling:	Functional
Protocol Version Control (PVC):	2 (if available; if not, select PVC 1)
TEI:	Dynamic
Number of Keys:	3
Release Indicator:	No
Ringing Indicator:	No
EKTS:	Off

B1.D. MOTOROLA BITSURFR TA210

Motorola Corporation
5000 Bradford Dr.
Huntsville, AL 35805-1993
Tel. 205-430-3000 Fax 205-830-5657

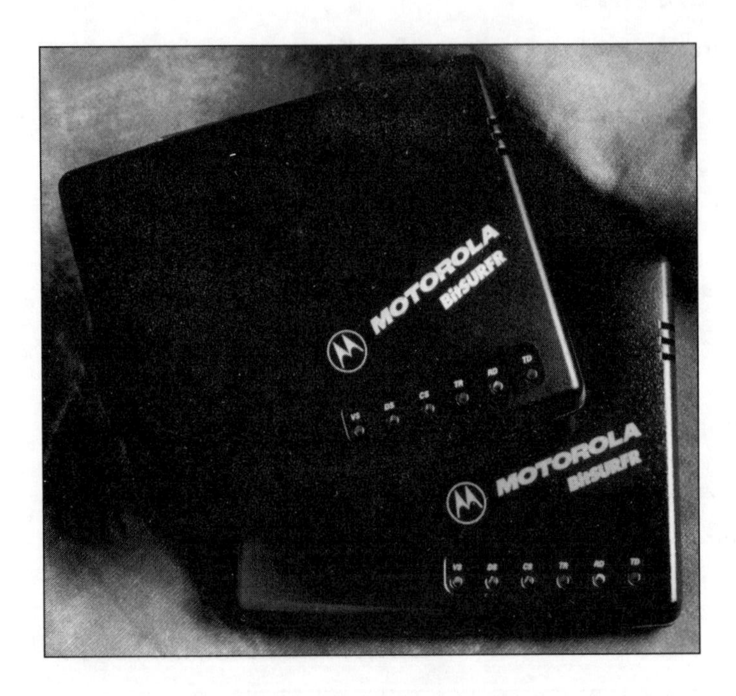

Motorola's BitSURFR is a Stand-Alone terminal adapter for high-speed ISDN telephone service. Ideal for the small office, branch office, or work-at-home user, BitSURFR is an all-in-one combination of hardware and software ready to connect your PC, telephone, and fax machine to ISDN.

Here's what you can do with BitSURFR and ISDN:

- Access all Intemet services at speeds of 64 Kbps
- Download digital video clips, music, speeches, and high-resolution color photographs from the World Wide Web and other Internet services

- Send and receive data files fast: desktop publishing documents, spreadsheets, engineering data
- Play interactive games

You can do all these things while talking on the telephone or sending a fax—at the same time you're sending or receiving data with your computer.

WORK AT HOME WITH BITSURFR AND ISDN

BitSURFR has the NT-1 interface built in, so no other equipment is required for full ISDN service. And you can connect to Microsoft® Windows Plug and Play™.

FEATURES

Data rates	300 to 115,200 bps asynchronous
	1200 to 64,000 bps synchronous
	Up to 128 Kbps with synchronous BONDING
Standards (signalling protocols)	National ISDN-1 Northern Telecom DMS-100 central office switch
	AT&T 5ESS central office switch
	Bellcore SR-NWT-001953
ISDN interface	U interface (2BlQ encoding per ANSI Tl.601)
Digital terminal interface (for your computer)	Standard EIA-232E,CCITT V.24
Phone interface	Standard RJ-11
Computer compatibility	PC with a 386 or 486 processor (or higher) and a 16550 or equiv alent high-speed serial port

Data compression	Provided in HyperAccess bundled software
B-channel protocols	V.120
	PPP (Point-to-Point Protocol)
	BONDING
Command set	Hayes® compatible AT commands
Size	6.4 in. wide, 5.3 in. deep, 1.5 in. high
Weight	10.5 oz

ISDN Line Ordering Information

The following is a list of ISDN line configuration requirements for the Motorola BitSURFR.

ISDN Physical Line Requirements

- 2B+D Basic Rate Interface, 2BlQ U interface (ANSI Tl.601)
- NIUF standard line configuration line set (NIIG) 11

ISDN Line Configuration Requirements:

The BitSURFR supports the following switch types and versions:

AT&T 5ESS	5E4.2 and later, custom or standard (NI-1)
Northern Telecom DMS100	BCS-29 and later, PVCO-PVCl or PVC2 (NI-1)
Siemens EWSD 10.0 and later	NI-1
Other NI-1 compliant	NI-1

The following bearer capabilities are required:

- Circuit mode voice (circuit-switched voice, CSV) service for speech and 3.lkHz audio

- Circuit mode data (circuit-switched data, CSD) for 56 Kbps and 64 Kbps unrestricted data
- Simultaneous circuit-switched voice (CSV) and circuit-switched data (CSD) calls

B1.E. 3COM/IMPACT

3COM Corporation
PO Box 58145
5400 Bayfront Plaza
Santa Clara, CA 95052-8145
Tel. 800-NET-3COM, 408-764-5000

The 3COM Impact™ ISDN External/digital modem provides ISDN connectivity at least four times the speed of analog modems. These products allow organizations to upgrade existing remote access equipment from analog to ISDN to deliver LAN-like performance. For SOHO users, 3COM impact is a useful tool for accessing resources on the Internet and the corporate LAN. 3COM Impact works with remote access server equipment as well as via its analog interface port.

FEATURES

- The 3COM Impact utilizes ISDN service available from the telephone company to transmit at 128 Kbps.
- The 3COM Impact always connects at the top speed available and maintains that speed for the entire call.
- For the remote user the 3COM Impact digital modem utilizes both ISDN B-channels, allowing simultaneous transmission of voice and data over a single ISDN line.
- 3COM Impact supports the standards-based communications software already used with modems through support of the AT command set and IETF Point-to-Point Protocol (PPP).
- V.120 rate adaptation or Async-Sync PPP™ conversion ensures interoperability with ISDN remote access equipment at the ISP or corporate location.
- 3COM Impact includes an integrated NT-1, eliminating the need for extra equipment.

ISDN LINE ORDERING INFORMATION

- Not Available

B2. ISDN LAN CARDS

B2.A. CISCO 740 SERIES ISA BUS ADAPTER CARD

Cisco Systems, Inc.
170 West Tasman Drive
San Jose, CA 95134-1706
Tel. 800-GO-CISCO

The Cisco 740 Series of ISA Bus PC adapter cards provide high performance ISDN access to enterprise LANs and Internet-based resources from individual PCs in the home or small office. With the 740 adapter, telecommuters and after-hours workers operate the same LAN desktop as in enterprise offices. They have identical access to all the shared applications and server-based files at the enterprise LAN. The Cisco 740 can also be used to access on-line bulletin boards and the Internet.

Plugging directly into the PC's ISA Bus, the Cisco 740 looks like an Ethernet adapter card to the PC's software. The 740 is also compatible with network operating systems, including Windows, Windows NT, Windows 95, Linux and Novell NetWare as well as TCP/IP-based applications.

The Cisco 740 Series is available in four models, with data compression and integrated NT-1 options. With the compression hardware, file transfer rates of up to 512 Kbps are achieved on the two aggregated B-channels of the ISDN basic rate line. All models have an S/T port for use with an ISDN phone, fax, or other ISDN devices.

FEATURES

- No Ethernet NIC cards or hubs required
- Data compression (on some models)
- Built-in NT-1 (on some models)
- On-demand dialing
- Supports ISDN phone, fax
- Compatible with all Cisco products
- Software upgradeable
- Authentication and call-back security
- Remote management
- ODI and NDIS drivers included

ISDN Line Ordering Information

Note: This information can be obtained from the manufacturer.

B2.B.1 DIGIBOARD PC IMAC

Digi International PC IMAC
6400 Flying Cloud Dr.
Minneapolis, MN 55344-9761
Tel. 800-344-4273 Fax 612-943-0404

Digi® IMAC™ is an ISDN LAN bridge PC board that provides a connection between Ethernet LANs and ISDN communications. The board connects directly to an Ethernet LAN cable and an ISDN BRI phone line using standard connectors. The IMAC supports two B-channels and one D-channel.

Digi IMAC technology supports any network operating system, LAN application, and ISDN CCITT/ANSI switch used by your telephone company. IMAC is designed to be compatible with Ethernet computing environments.

FEATURES

- Industry standard data compression
- Transparent connection management; no dialing commands necessary
- Supports any LAN application software
- English language configuration and control
- Programmable operation
- Context-sensitive help screens
- Remote control operation over ISDN or Ethernet
- Automatic security feature based on incoming call ID
- Automatic fallback from circuit-switched 64 Kbps calls to receive 56 Kbps calls or send at 56 Kbps over ISDN voice calls
- Built-in Ethernet and ISDN protocol analyzer
- Complete trouble shooting messages available
- Network management
- 256K flash ROM for easy updating of new features as they become available
- SNMP MIB II
- IMAC supports two B-channels and one D-channel
- Supports "nailed-up" permanent circuit-switched B-channel
- Compatible with 64 Kbps, 56 Kbps, and ISDN voice circuit-switched calls
- Supports Q.931 and Q.921 (LAPD), HDLC, and proprietary protocols on voice and circuit-switched data channels
- Compatible with all CCITT/ANSI BRI ISDN central office switches, including AT&T 5E, Northern Telecom DMS-100, Siemens Stromberg-Carlson, British Telecom, and others

- Supports single-point and multipoint (passive bus) line configurations
- Connects between ISDN switches using digital dial-up voice lines and digital trunking

Ethernet Bridge

- Supports all Ethernet network operating systems
- Ethernet remote address filter table—up to 2047 entries—filter on first 64 bytes of Ethernet packet—automatic or selectable address table—fixed address table
- Adaptive address learning algorithm
- Filtering rate: 14,400 Ethernet packets per second
- Forwarding rates, IMAC: 250 packets per second over two B-channels
- Manual accept/reject mode addresses accepted or rejected according to selection

I/O Interfaces

- ISDN S/T Basic Rate Interfaces via RJ-45 connection
- RS-232 asynchronous serial port for local configuration and control
- IEEE 8023 BNC and AUI switch-selectable connectors

ISDN Line Ordering Information

1. Determine the switch type you will be connecting with. Based on your phone number, the telephone company should be able to tell you the type of switch located in your area; you most likely will connect with one of these three switches: AT&T 5ESS, Northern Telecom DMS-100, or National ISDN-1 (NI-1).
2. You may need to provide the phone company with the switch translation best suited for the type of switch you will be connecting with. Refer to the table below:

	AT&T	Northern Telecom	National ISDN-1
Protocol	5ESS	DMS-100	
	Proprietary or NI-1	Functional or EKTS-No	NI-1
Supplementary Features	Terminal Type A or D	EKTS=No	EKTS=None*** (Recommended)
Multipoint	Yes (or point to point)	Yes	Yes
Bl Protocol	Circuit Switched Data and/or Voice	Circuit Switched Data and/or Voice	Circuit Switched Data and/or Voice
B2 Protocol	Circuit Switched Data and/or Voice	Circuit Switched Data and/or Voice	Circuit Switched Data and/or Voice
# of SPIDs *	None, 1 or 2**	2	2

*SPIDs = Service Profile Identifiers

**Two SPIDs are required to place simultaneous voice 56 call (on both B-channels of an ISDN BRI line)

***EKTS = basic may be used, but if EKTS = basic, you will not be able to place two B-channels calls using just one logical device.

3. Be sure to ask the phone company which protocol your switch is running.

4. If you are connecting to a DMS-100, verify that EKTS is disabled. For NI-1 switches, EKTS should be set to basic or none.

5. Choose a multipoint connection if:

   - You are connecting with an AT&T 5ESS running proprietary protocol and want to make two simultaneous voice channel calls.

   - You plan to attach two or more devices to a single NT-l. Otherwise, choose point-to-point service.

6. Be sure the phone company gives you a SPID number for every logical terminal on a multipoint line.

- AT&T 5ESS supports up to eight SPIDs per line.
- Northern Telecom DMS-100 and National ISDN-1 support two SPIDs per line, one for each B-channel.

7. You will also need the phone number and dialing pattern assigned to each logical device. ISDN phone numbers can be as short as three numbers or more than seven numbers.

B2.B.2. DIGIBOARD DATAFIRE

Digi International DataFire
6400 Flying Cloud Drive
Minneapolis, MN 55344-9761
Tel. 800-344-4273 Fax 612-943-0404

Digi® DataFire™ is an ISDN terminal adapter/network interface PC card that incorporates the NT-1 U interface feature, providing remote LAN connections for individual computer users. The DataFire installs in an expansion slot of ISA-bus PCs and connects directly to an ISDN Basic Rate Interface (BRI) phone line.

One 128 Kbps or two 64 Kbps communications channels are supported. DataFire integrates with Microsoft® Windows NT™ Windows for Workgroups and Novell® NetWare® and DigiBoard's Line Optimizing Software. DataFire supports PPP (Point-to-Point Protocol) to provide interoperability with other types of ISDN devices, facilitating connections with the Internet.

FEATURES

- Supports two B-channels and one D-channel; B-channel voice or circuit-switched data connections
- Supports "nailed-up" permanent circuit-switched B-channel
- Compatible with 64 Kbps, 56 Kbps, and ISDN voice circuit-switched calls
- Supports Q.931 and Q.921 (LAPD), HDLC, and proprietary protocols on voice and circuit-switched data channels
- Compatible with all CCITT/ANSI BRI ISDN central office switches, including AT&T 5E, Northern Telecom DMS-100, Siemens, British Telecom, and others
- Connects between ISDN switches using digital dial-up voice lines and digital trunking
- Data compression for greater throughput
- Point-to-Point Protocol (PPP) support for interoperability
- Transparent connection management; no dialing commands necessary
- Programmable operation

- Automatic fallback from circuit-switched 64 Kbps calls to receive 56 Kbps calls or send at 56 Kbps over ISDN voice calls

I/O Interfaces

- ISDN U interface accepts RJ-45 or RJ-ll connectors

Network Interface

- Supports client LAN software for Novell, Microsoft RAS, and Internet PPP server types
- Supports DOS ODI 3.11 and DOS NDIS 2.0
- Frame rate: 230 packets per second over two B-channels
- 8-byte I/O port, seven available addresses
- 1 software vector, 60H-80H
- Interrupt vector (IRQ) 3, 5, 7, 10, 11, 12, 15—can be run without IRQ
- No memory required
- 8- or 16-bit slot

ISDN Line Ordering Information

Follow the steps listed for ordering ISDN lines with the DigiBoard PC IMAC, on pages 234–236.

B2.C. U.S. Robotics Sportster ISDN 128K™

U.S. Robotics
8320 Old Courthouse Rd.
Suite #200
Vienna, VA 22182
Tel. 703-883-0933 Fax 703-883-8043

Sportster ISDN 128K™ is an ISDN terminal adapter PC card that installs in an expansion slot in your PC and provides the interface with an ISDN BRI line. Sportster ISDN 128K's ISDN network connection allows users to send and receive information to and from the Internet, on-line services, and other sources.

Sportster ISDN 128K has NDIS, ODI for (IPX), packet driver, and TAPI interfaces so that it can be used with off-the-shelf software packages to access the full resources and services of remote locations.

The Sportster ISDN 128K was designed to handle data and voice. Sportster ISDN 128K is equipped with an RJ-l l jack so that you can plug in an analog telephone to communicate over the digital network. ISDN allows you to simultaneously use the voice and data communications capability of this product.

Dynamic bandwidth allocation, compression, PPP, multilink PPP, integrated NT-1, RJ-l l analog phone jack, Intemet access, security via Secure-ID card (in conjunction w/SecureLink Server), and TAPI provide a number of communications applications.

FEATURES

- Requires PC 386 CPU or higher, 4 MB RAM, DOS 5.0 or higher
- MS-DOS, Windows, and Windows for Workgroups
- NDIS, ODI, packet and TAPI drivers
- Integrated NT-1
- Compression
- PPP and Multilink PPP for full interoperability with other vendors' equipment
- ISDN National-1, AT&T Custom and NT-1 Custom, CCITT standards
- Analog phone support with voice input and keypad dialing

- Security via Secur-ID card (in conjunction w/SecureLink Server), MAC address and caller ID
- Full support of TCP/IP, IPX, NETB1OS, DECnet, and other protocols over ISDN
- Dual B-channels for high-speed data and voice
- Internet Access
- Simultaneous voice and data
- Software Configurable (no Jumpers)

ISDN Line Ordering Information

US Robotics recommends requesting Intel Blue ISDN service, which is compatible with Sportster, from the local exchange carrier.

B2.D. ISDN*TEK

ISDN*tek
PO Box 3000
San Gregorio, CA 94074
Tel. 415-712-3000 Fax 4l5-712-3003

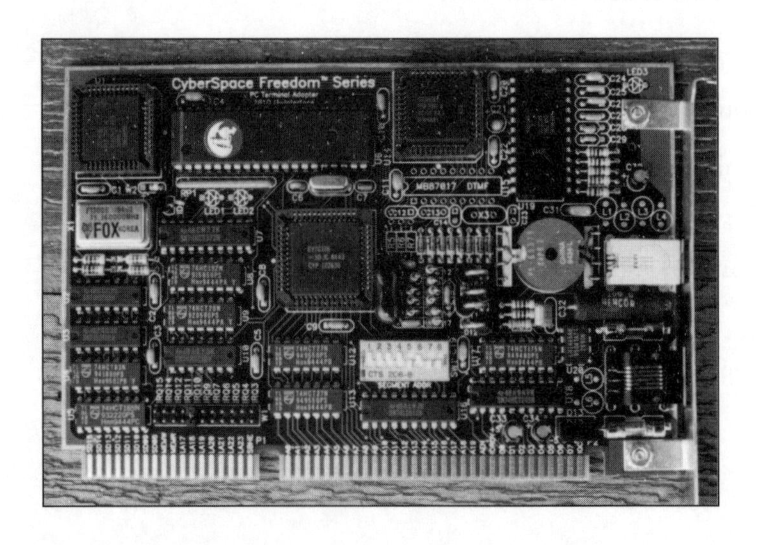

CyberSpace Freedom Series

The CyberSpace Freedom Series was designed to provide a low-cost hardware connection to the global ISDN network supplied by the telephone companies. All of the CyberSpace cards can interface to the Internet via appropriate third-party TCP/IP-PPP software and an Intemet access point. They may also connect peer-to-peer to any other TCP/IP equipped computer, using worldwide ISDN for fast, reliable communications.

The half-size PC card slips into your (E)ISA PC-bus and includes the phone cable to connect directly to the NT-1 or jack on your wall. The card is provided with "WinISDN" driver software on 3.5" diskette and with installation and diagnostic software that runs on Windows PCs. The CyberSpace Freedom™ Card supports virtually all Basic Rate ISDN found in the U.S. today.

Features

- All CyberSpace products feature automatic switch detection; the hardware needs no special configuration to work with the five varieties of telco switches.
- The CyberSpace cards can both place and answer calls. Coupled with software such as Netmanage Chameleon, the cards operate as either client, server, or peer.
- Because of its hardware-based protocols, the CyberSpace card provides the fastest connect time possible with ISDN.
- Interoperability testing has shown the CyberSpace card to be among the fastest in 128K throughput for high speed data transfers.

The Hardware

- PC-based interface to ISDN.
- Low profile, half-size PC-card.
- IBM PC/AT ISA or EISA bus interface.

- Jumper selectable interrupts (IRQ-3...IRQ-15).
- Switch selection of PC-bus addresses.
- 4K byte on-board memory can be any block in 16Mbyte address space.
- Setup program assists in installation, switch setting, and troubleshooting.
- FCC part 15 class B certified for home or office use.

The ISDN Connection

- Auto-detects and works with virtually any U.S. ISDN Basic Rate line available.
- Integral ITU (CCITT) Q.931, Q.921, and 1.451 support.
- Approved by Northern Telecom and Bellcore.
- Basic Rate ISDN interface (2B+D, 2B, 1B, or 2B+s).
- Preset 100 ohm line termination resistors are jumper removable.
- Diagnostics report ISDN Layer 3 messages to user.
- Supports 56Kbits/sec to 64Kbits/sec data call on either B-channel.
- Voice product additionally supports a single voice call on either B-channel.
- Two-channel data product is 128K-ready for MultiLink PPP (MP) software.
- "Standard" series uses 8-wire RJ-45 connector to S/T jack of external NT-1.
- "Plus" series uses 6-wire RJ-ll connector to U jack using onboard NT-1.
- Voice product uses 4-wire handset connector for headset or handset.
- "Plus" series 2BlQ U interface conforms to ANSI Tl.601-1988 at 18,000 ft maximum distance.

- "Standard" series S/T interface conforms to ITU/CCITT I.430 at 1,000 ft maximum distance for a single device on the ISDN line.

The Driver Software

- The WinISDN.DLL driver supports synchronous PPP (Point-to-Point Protocol) and HDLC from the Windows-based TCP/IP stacks below.
- Works with Service Provider's ISDN routers when used with any WinISDN-based PPP software.
- Supports Netscape, FTP, Ping, e-mail, Gopher, Mosaic, etc., when used with appropriate TCP/IP software.
- Driver support for Peer-to-peer connectivity, messaging, file transfers, streaming data, and voice connections.
- Voice product includes dialer/phonebook mini-app.
- Easy interface to driver from Visual Basic or C programs.
- Software Developer's Kit available.

Compatible Software Running on ISDN*tek Hardware

- Netmanage Chameleon NFS and Internet Chameleon
- Spry Internet in a Box
- FTP Explore OnNet
- Frontier Technologies' SuperHighway Access
- Shiva Corporation's ShivaPPP for Remote LAN Access

Compatible Hardware at Other End of the ISDN Connection

- 3COM Impact
- Adtran

- Ascend (Pipeline and MAX)
- Cisco
- CoSystems
- DigiBoard (PC/IMAC and DataFire)
- Eicon/Diehl
- Farallon Computing
- Flowpoint
- Gandalf
- IBM WaveRunner
- ISDN*tek
- JRL Systems
- KNX
- Microsoft
- Motorola
- MPX
- Network Express
- Novell
- Shiva Corp's ShivaPPP and LANRover
- Skyline Technologies
- Stagecoach
- TeleSoft
- US Robotics/ISDN Systems
- Xyplex Network 3000

ISDN Line Ordering Information

Depending on which regional phone company is serving you, you will probably find that your choice of line configurations is an issue of economics. We recommend that you get as many features as your budget allows. You can always add more features later as you need them.

NIUF Capability Codes

Last summer the National ISDN User's Forum (NIUF) got together and drew up some guidelines for defining the more popular line configurations. These attempt to limit the choices to a few dozen capability sets. We have chosen the NIUF ISDN Capability Sets that seem most logical for the CyberSpace products, and we recommend a few that will give you the most flexibility.

NIUF Capability codes for ordering ISDN.

NIUF Code	Cyber Card	# of Chan	Chan Type	Typical Use Description
B.	I	(lB)	CSD	64K Internet access (data only)
C.	I,C	(lB)	CSVD	64K Intemet OR voice
G.	I,C	(2B)	CSD+CSV	64K Internet AND voice (best choice for 64K Internet and POTS)
I.	I	(2B)	CSD+CSD	64K or 128K Internet (data only)
J.	I,C	(2B)	CSD+CSVD	64K Internet and voice, OR 128K Internet (Good choice for 128K Internet and sporadic POTS)
K.	I,C	(2B)	CSD+CSVD	same as J but with some calling features that are not used by the CyberSpace cards.
L.	I,C	(2B)	CSD+CSVD	same as K but with EKTS for an ISDN telephone set with programmable feature keys.

Continued

NIUF Code	Cyber Card	# of Chan	Chan Type	Typical Use Description
M.	IC	(2B)	CSVD+CSVD	64K data and Voice, OR 128K data, OR two voice lines (best choice for the Commuter Card and the most flexible choice for the Internet Card with POTS)

"CyberSpace Card" Line Configuration Template

AT&T 5ESS	Custom	NI-1 Custom
(B-channels for CSV and CSD)		
(D-channel for signalling only)		
# channels for CSV	2	2
# channels for CSD	2	2
Terminal Type	A or D	A or D
# of Call Appear	N/A	1
Display (Y/N)	N/A(Y)	N/A(Y)
Prefer Ringing/Idle	N/A	I
Autohold (Y/N)	N/A	N
Onetouch (Y/N)	N/A	N
EKTS	OFF	OFF
Multipoint	YES	NO

DMS-100	NI-1	Custom
(Voice and Data for each B-channel)		
(No D-channel packets)		
Functional Signalling	Y	Y
PVC Protocol Version	2	1
Dynamic TEI	Y	Y

DMS-100	NI-1	Custom
Max# prgrmable keys	N/A(3)	N/A(3)
Release Key (N/Key#)	N	N
Ringing Indicator (Y/N)	N	N
EKTS (Y/N)	N	N
CACH (Y/N)	N	N

B2.E. PLANET-ISDN BOARD

Tel. 408-446-8690 Fax 408-446-9766

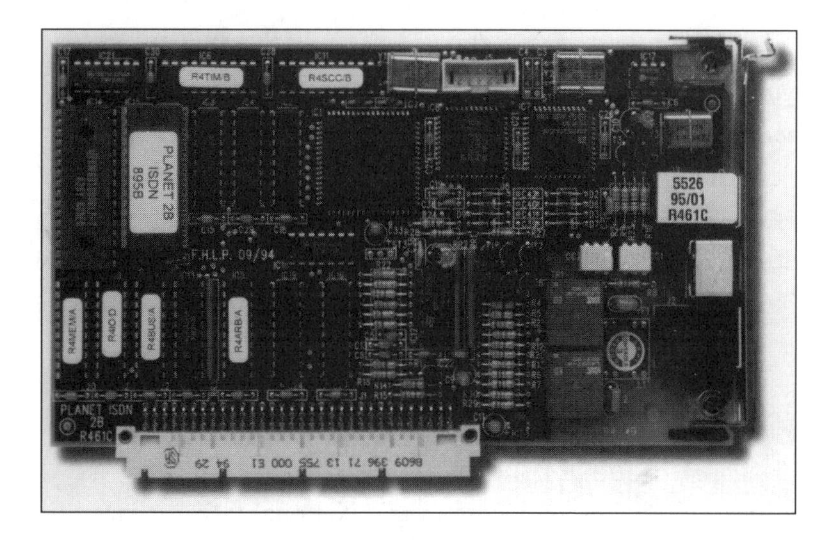

Compact 2B+D ISDN NuBus Board for Macintosh

The Planet-ISDN communication board connects your Macintosh to ISDN. You can print a file at a remote site or use your Macintosh for videoconferencing with customers, suppliers, or coworkers. If high-speed Internet connectivity is your need, use the Euronis PPP driver to set up a synchronous PPP connection to a host Internet service provider. Planet-ISDN (also known as the Planet-ISDN II Board) enables your Macintosh to take advantage of the many networking features (both data and voice) provided worldwide through ISDN.

Features provided by Planet-ISDN can be used simultaneously for a wide range of applications such as high-speed data transfer (files, pictures, sound, text, etc.), connection to remote LANs and BBSs, remote maintenance, videoconferencing, and Internet access.

Working as a background task on your Macintosh, the Planet-ISDN Phone is a telephony software application that manages incoming and outgoing voice calls on one of the B-channels. Among its features you'll find a built-in directory.

Planet-ISDN is available and approved for use in Australia, Belgium, Canada, Denmark, Finland, France, Germany, Hong Kong, Ireland, Italy, Japan, Netherlands, New Zealand, Norway, Portugal, Singapore, Spain, Sweden, Switzerland, the United Kingdom, and the United States. In the United States, the Planet Board works on ISDN lines originating from AT&T 5ESS, Northern Telecom DMS-100, or Siemens EWSD central office switches (both custom and National ISDN-1 lines).

Planet-ISDN can be installed in practically any Macintosh with a NuBus slot (even in the smaller slot computers, such as the Quadra 610).

FEATURES

- 2B+D board, working on both custom and National ISDN-1 lines
- 2B-channels, which can be used for data only or for voice and data
- Small 7" board, highly integrated, which can fit into smaller Macs such as the Quadra 610 or PowerMac 6100
- Native software for both 68K and PowerPC Macs
- RJ-45 port for ISDN
- RJ-11 port for analog service (POTS, modem, fax, etc.)
- Protocols:

    - HDLC
    - X.25 64 Kbps
    - X.25 128 Kbps (2 B-channels)
    - Multi-B-channel X.25 for nx64 connections
    - Synchronous PPP for Internet connections
- Can put one to six boards in a single Mac
- Supports 56 Kbps rate adaptation
- CommToolBox compatible (can use existing commercial software)
- Incoming call filter so that you can have multiple applications waiting on an incoming call
- Incoming calls routed to proper applications and/or sessions

The Planet-ISDN Board uses an external network terminator (NT-1) so other ISDN devices can be used on the same ISDN line (NI-1 only).

Software

Euronis also offers some excellent software, including EasyTransfer (a high-speed file transfer application, which is also supported on the Euronis Gazel Board for Windows PCs) and The Link, an AppleTalk Internet router software package for LAN-to-LAN connections. The EasyTransfer software is easy to set up. Simply drag and drop a file or folder over to an appropriate Correspondent icon, and the connection will be made, the file and/or folder will be transferred, and the connection will be dropped. You can also use EasyTransfer in its full foreground mode to retrieve files and/or folders from a remote machine. EasyTransfer also supports cross-platform transfers (Mac to PC and PC to Mac) and multiple B-channel connections (for up to 25 Kbps transfers). EasyTransfer is used in the graphics and prepress industry for transfer of multi megabyte files. You can expect file transfer speeds of 1MB per minute with a single Planet-ISDN board and EasyTransfer.

Due to its support of the Communications Tool Box through the Planet-ISDN Tool, other commercial software such as AppleTalk Remote Access (using the Diplomate software), Timbuktu, FirstClass, and Telefinder e-mail—Bulletin Board Systems, Imagexpo (prepress screen layout conferencing software), and others easily work with the Planet-ISDN Board.

Internet and AppleTalk Connectivity

For high-speed Internet connections, the Planet-ISDN board comes bundled with EuronisPPP. EuronisPPP is a software application and set of extensions that enables you to make high-speed connections to commercial Internet service providers (single B-channel). The EuronisPPP application supports both PAP and CHAP security options, and it is also able to make AppleTalk connections to routers that support the ATCP protocol (AppleTalk over PPP). Once a EuronisPPP link is established, you are able to use IP-based programs (Mosaic, Netscape, Eudora, Fetch, etc.) over this link. If your router also supports the ATCP protocol, you will also be able to mount remote volumes, use file sharing, print to remote printers, and use any remote AppleTalk network resource.

The Planet-ISDN board supports videoconferencing/groupware applications through software and hardware packages produced by SAT (Meet-Me) and Intelligence at Large (Being There).

ISDN Line Ordering Information

Planet ISDN works with all the switch types (AT&T, Northern Telecom, and Siemens). The ISDN lines offered by your phone company may differ from region to region. You may be given an option of a custom line or National ISDN-1. If you are given a choice, order the National ISDN-1 configuration. Follow these guidelines when ordering the specific type you choose.

AT&T Custom 5ESS (5E6, 5E7, 5E8, and 5E9)

Service types on both B-channels are as follows:

- Circuit-switched data/voice (CSD/V)

- Circuit-switched data (CSD)

(You can choose to have voice capabilities or not.)

Phone numbers:	Single*
Line configuration:	Point-to-point (very important)
Missing	
Max. number of B-channels:	2
Number of call appearances:	1 if you order voice (CSD/V), otherwise none
Terminal type:	D if you order voice (CSD/V), otherwise A (basic call terminal)
Bearer service restrictions:	None (DMD on both channels)
EKTS:	No requirement
One Touch:	No
Autohold:	No

Northern Telecom DMS Custom

Service types on both B-channels are as follows:

- Circuit-switched data/voice (CSD/V)
- Circuit-switched data (CSD)

(You can choose to have voice capabilities or not.)

TEI assignment:	Dynamic
Telephone numbers:	2 (with SPIDs)
Terminal type:	A (Basic call terminal)
Bearer service restrictions:	None
EKTS:	No requirement

Ringer indicator:	Yes
Authorized call types	CMD
EKTS:	No requirement

B3. ISDN ROUTERS AND BRIDGES

B3.A.1. ASCEND PIPELINE 25

Ascend Communications Corporation
1275 Harbor Bay Blvd.
Alameda, CA 94502
Tel. 510-769-6001 Fax 510-814-2300

The Ascend Pipeline 25 ISDN connects computers in small offices and homes to backbone networks or the Internet over ISDN lines. The Pipeline 25 can also connect two analog devices over the same ISDN line. The Pipeline 25 provides access to corporate network resources or the Internet while combining telecommunications needs into one line.

In a single modem-sized box, the Pipeline 25 ISDN combines:

- Bandwidth on demand
- ISDN Basic Rate (BR) terminal adapter
- Bridging and optional routing
- LAN and WAN network management
- Optional data compression
- Dual integrated analog interfaces
- Built-in ISDN NT-1
- Security

Bandwidth on Demand

Dial-up connections are automatically established and removed as needed.

- Inverse multiplexing uses both ISDN B-channels for 112/128 Kbps data rate.
- Dynamic bandwidth allocation varies bandwidth.
- Bandwidth supports two simultaneous connections to different locations.

Dual Integrated Analog Interfaces

Dual interfaces allow consolidation of lines.

- Dual interfaces connect a combination of telephones, modems, and fax machines.
- Sophisticated call routing directs calls to the correct analog device.
- Incoming and outgoing calls preempt one B-channel while maintaining one B-channel for data connections.
- Both analog devices can be used while data is idle.

Integrated ISDN BRI

This offers a fully integrated ISDN interface for plug-and-play operation.

- ISDN BRI
- Standard S/T or U interface (national ISDN-1 compliant), which eliminates need for external NT-1 device
- Pipeline 25 (ISDN), which provides high-speed, switched digital access to the Internet over ISDN BRI

Access Routing and Bridging

Advanced protocol support ensures efficient connectivity to all LANs and the Internet.

- Standard multiprotocol bridging

- Optional IP or IPX routing
- PPP, Multilink PPP, and MPP (Ascend's Multichannel Point-to Point protocol)
- Transmit and receive packet filtering

LAN/WAN Management and Control

Manage your remote access equipment and bandwidth as easily as you manage your local LAN facilities with the following:

- Rigid security features
- Call detail reporting—WAN loopbacks
- Flash memory for software downline loading

Security

The Ascend Pipeline 25 employs rigid security features that ensure network access only to authorized users with the following:

- PAP, CHAP, callback, calling number ID, password, TACACS
- Token-based security including Secure-ID and Enigma

ISDN LINE ORDERING INFORMATION

AT&T 5ESS—National ISDN-1

Request from the telephone company a National ISDN-1 ISDN line in a multipoint configuration with 2BlQ line code. The multipoint configuration will allow you to have a separate telephone number for each B-channel; however, it will physically be only one ISDN line. The telephone company should supply you with a different telephone number and SPID for each B-channel. The SPID format is 01 +7-digit telephone number + 000 (01XXXXXXX000). Your ISDN line must be configured to allow voice and circuit-switched data on both B-channels and signalling on the D-channel. Request that the telephone company program your ISDN line with the following attributes:

- Maximum terminals set to 2 (This tells the switch that there are two terminals active on this line.)
- Maximum B-channels set to 2; Actual User set to Yes (This tells the switch that you are an actual user and may use both B-channels simultaneously.)
- Circuit-switch voice set to 1; circuit-switch voice channel set to Any (The switch only allows 1B-channel to actually be active for voice at a time. The Any tells the switch that it can use either B-channel to deliver the call.)
- Circuit-switched data set to 2; circuit-switched data channel set to Any (This tells the switch that you may connect both B-channels simultaneously. The Any tells the switch that either B-channel may be used for data.)
- Terminal type is Type A - Basic Terminal (AT&T has defined the terminal types by letters. This tells the switch that you are a basic National ISDN-1 terminal.)
- Display set to Yes (This tells the switch that you have display capabilities.)
- Circuit-switch voice limit set to I (This tells the switch that you may receive up to one voice call.)
- Circuit-switched data limit set to 2 (This tells the switch that you may receive up two data calls.)

The telephone company will also need to know any additional voice features that you require on your ISDN lines. Examples of these features are caller ID and call forwarding, call hold, flexible calling, etc.

DMS-100 BCS-35 National ISDN-1

Request from the telephone company a National ISDN-1 ISDN line with 2BlQ line code. Your ISDN line must be configured to allow voice and circuit-switched on both B-channels and signalling on the D-channel. Normally you will use the Bl-channel for voice calls and the B2-channel for data calls. The telephone company should supply you with a different telephone number and SPID for each B-channel. Request that the telephone company program your ISDN line with the following attributes for B1 and B2:

- Set the circuit-switch option to Yes; set the bearer restriction option to no packet mode data (NOPMD) only (This tells the switch that you require circuit-switch ability on your B-channel. The bearer restriction on your line means that you are not allowed to do packet data on your B-channel.)
- Set protocol to functional version 2 (PVC 2) (This tells the switch that your CPE software is using the National ISDN-1 protocol.)
- Set the service profile identification (SPID) suffix to 1 [This tells the switch that the digit following your telephone number will be 1. The SPID format is area code + 7-digit telephone number + 1 + 00 (XXXXXXXXXX100.)]
- Set the terminal endpoint identifier (TEI) to Dynamic [This tells the switch that you can accept any TEI value from 64 to 126. The assignment of a dynamic TEI is the responsibility of the switch.]
- Set ring to Yes (This tells the switch to send an alerting message to your CPE when there is an incoming call.)
- Set the maximum keys to 10 (This tells the switch how much memory to allocate for features.)
- Set key system (EKTS) option to No (This tells the switch that you are not a key system. A key system is where multiple telephone numbers are shared across terminals.)
- Place the lower layer compatibility option for data on this B-channel. (This tells the switch that your CPE may utilize the lower layer compatibility information element for compatibility checking with the called CPE.)

The telephone company will also need to know any additional voice or data features that you require on your ISDN lines. Examples of these features are caller ID, call forwarding, call hold, flexible calling, etc.

B3.A.2 ASCEND PIPELINE 50

Ascend Communications Corporation
1275 Harbor Bay Blvd.
Alameda, CA 94502
Tel. 510-769-6001 Fax 510-814-2300

The Ascend Pipeline 50 is an external Ethernet-to-ISDN router/bridge. Telecommunications capabilities and inverse multiplexing are combined with standard-based bridging and routing. Remote PC-to-WAN/LAN, LAN-to WAN/LAN connectivity can all be done. The unit connects multiple users on an Ethemet network through a single ISDN connection.

When integrating the hardware the user connects one of the existing hub's 10BaseT ports to the Pipeline 50 LAN port, then connects the Pipeline 50 WAN port to the ISDN line.

Using Multichannel Point-to-Point protocol (MPP), the Pipeline 50 uses both B-channels simultaneously to increase available bandwidth to 128 Kbps. The hardware data-compression scheme is 2:1, supporting speeds up to 256 Kbps.

Five status LEDs (power, activity, collision, WAN, and condition) are visible on the exterior of the unit. The management interface is accessible through a serial modem cable and communications software. The interface includes statistical displays (current session, telephone line, and LAN and WAN connections) and support for SNMP (Simple Network Management Protocol).

Some of the main features of the Pipeline 50 are its bridging and routing capabilities, including simultaneous bridging and routing to multiple destinations, support for inverse multiplexing, user-definable dynamic bandwidth allocation, and user-definable filter sets.

Standard routing protocols are supported: Internet Protocol (IP), Point-to-Point Protocol (PPP), Multilink Protocol (MP), and Multichannel Point-to-Point Protocol (MPP). Telnet session capability is possible at local and remote locations.

The entire package includes the Pipeline 50 bridge/router, power transformer, DB-9 male to DB-25 female adapter, RJ-48C straight-through cable for l0BaseT, two manuals (Getting Started and Advanced Features), and a straight-through cable for the UI/F network connection.

The Pipeline 50 HX, which is a single-user variant of the Pipeline 50, supports all of the above with the exception that it connects only one computer to the WAN/remote LAN.

B3.B. CISCO 750 SERIES ETHERNET TO ISDN ROUTER

Cisco Systems, Inc.
170 West Tasman Dr.
San Jose, CA 95134-1706
Tel. 800-GO-CISCO

The Cisco 750 family of multi-protocol routers provides single users, small offices, and branch offices with high-speed remote access to enterprise networks and to the Internet. With the Cisco 750, telecommuters and remote office workers using an Ethernet workstation, PC, or Macintosh operate the same LAN desktop as in enterprise offices. Subject to authentication and customized access control via Cisco's Personal Network Profiles, these users gain remote access to server-based information and shared applications. They can also connect to the Internet or make temporary dial connections for office-to-office digital package delivery.

The Cisco 750 employs 4:1 compression hardware and an ISDN Basic Rate Interface (BRI) for speeds up to 512 Kbps. All models include an Ethernet port for LAN data. The Cisco 752 models also include an S/T port for

ISDN phone and fax support, and the Cisco 753 model includes an analog port for analog phone and fax support.

The Cisco 750 family offers IP and IPX routing, concurrent bridging, Multilink Point-to-Point protocol (PPP), SNMP management and multi-level-security. The Cisco 750 is available in six configurations: the 751, 752, and 753 SOHO models support up to four users, while the unrestricted 751, 752, and 753 RO models support remote offices and the enterprise. An optional internal NT-1 is available on both the SOHO and RO models.

Features

- Full function bridge-router for IP and IPX networks
- Multilink PPP
- Split B-channels
- On-demand dialing
- High performance data compression
- Remote management via SNMP and Telnet
- Link authentication (PAP/CHAP) and call-back security
- Personal Network Profiles
- TACACS and token-based authentication (with Connection Manager)
- Built-in NT-1 option
- ISDN phone, fax support
- Optional analog port for standard telephone modem or fax
- Software upgradeable
- Compatible with all Cisco products
- Worldwide certification

ISDN Line Ordering Information

Note: This information can be obtained from the manufacturer.

B3.C. GANDALF 52421

Gandalf Systems Corp.
501 Delran Pkwy.
Delran, NJ 08075
Tel. 609-461-8100 Fax 609-461-4074

XpressConnect™ LANLine 5242i is an integrated ISDN terminal adapter with NT-l. LANLine 52421 plugs directly into the ISDN BRI wall jack and features touch-tone telephone configuration.

LANLine 5242i dynamically allocates bandwidth to data and voice on the ISDN B-channels. Another LANLine 5242i standard feature is Gandalf's industry leading patented data compression. Gandalf's data compression technology minimizes file transfer times and lets remote users experience the same response times they would if they were using network resources locally.

LANLine 5242i's voice channel supports an analog phone, fax machine, and/or answering machine. The availability of two internetworking software packages lets you choose the feature set that best satisfies the distinct needs of your network and application. You can choose between 5242i Teleworker Bridge software or 5242i Edge Router software.

LANLine 5242i Teleworker Bridge software features user transparent dial-on-demand for IP address, IPX address, or any LAN traffic. Teleworker Bridge software also supports spoofing to minimize unnecessary traffic over the ISDN link. Security is maintained with calling device identification through LANLine 5242i's serial number.

LANLine 5242i Edge Router Software allows you to configure an IP (Internet protocol) route for enhanced security with network segmentation. For interoperability with other vendors' communications equipment, Edge Router software supports PPP (Point-to-Point Protocol), Multilink PPP, and CHAP and PAP security features. In addition to Gandalf's 4:1 data compression technology, 5242i with Edge Router software also features 4:1 Stacker L25™ Compression.

MS-Windows™ based console is also included on 5242i IP Edge Router software.

Both Teleworker Bridge software and Edge Router software provide access for quick network management from any remote location.

The XpressConnect LANLine 5242i belongs to a family of remote access, concentration, and internetworking products available from Gandalf.

Features

- IP, IPX, or any LAN traffic dial-on-demand over ISDN
- Quick connect/disconnect button
- Up to 8:1 data compression
- Data and voice bandwidth-on-demand
- Analog voice support
- ISDN S and U interface support
- PPP (Edge Router only)
- MS-Windows™ console (Edge Router only)

ISDN Line Ordering Information

When you order the connection from the ISDN service provider, you must specify your needs exactly. This appendix explains what you should ask for when ordering an ISDN line for the LANLine 5242i.

With an ISDN line, you can use your dial tone telephone, modem, fax, and answering machine in conjunction with the 5242i to place voice and data calls to remote locations.

North America

To support incoming call bumping, the ISDN switch must provide the incoming call to the LANLine 5242i, even if both B-channels are busy. This feature is usually a supplementary service at extra cost and is generically called ACO (additional call offering) or call waiting. You must request ACO, regardless of switch type. Outgoing call bumping requires no special ISDN service.

National ISDN-1 (NI-1)

For this switch type, you should request capability K. This ordering code includes alternate voice/circuit-switched data on one B-channel, and circuit-switched data on the other B-channel. This package provides voice features, including flexible calling, additional call offering, and calling number identification. Data capabilities include calling number identification.

B4. Network Termination Devices (NT-IS)

B4.A. Adtran

Adtran Inc.
Sales and Marketing Dept.
901 Explorer Blvd.
Huntsville, AL 35806-2807
Tel. 800-973-8726 Fax 205-971-8699

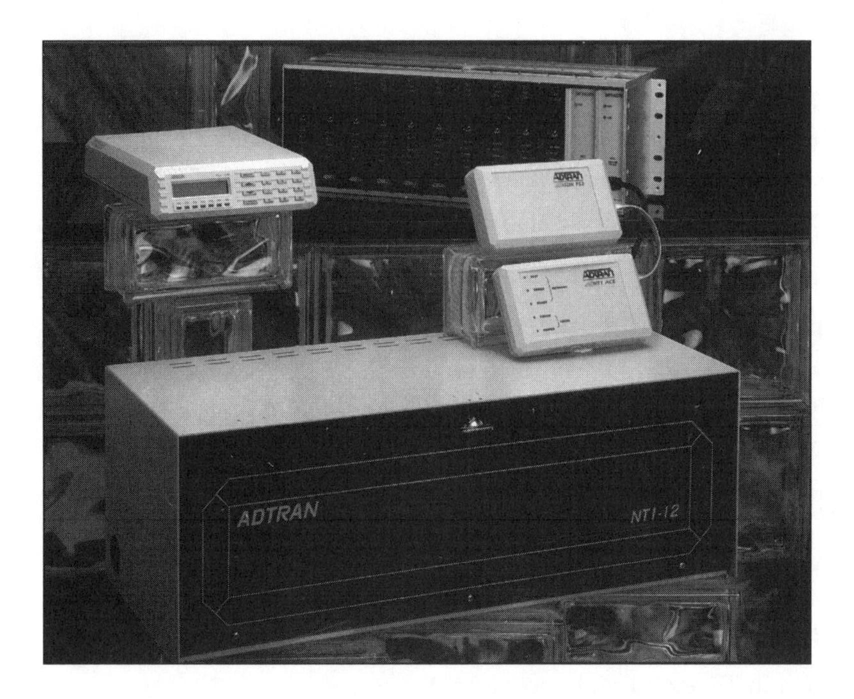

Adtran's Type 400 NT-1 provides the network termination for 2BlQ Basic Rate Interface between the customer's terminal equipment and the ISDN network. The U interface is used for connection to the ISDN network. Connection to the subscriber's terminal equipment is made via the S/T interface. The U and S/T interfaces conform to ANSI and CCITT standards and perform all layer 1 functions. The NT-1 contains lightning protection circuitry to protect the unit from lightning and line transients on the network interface. Five LED status indicators on the front panel of the unit aid in fault isolation.

The NT-1 is designed to operate as a stand-alone unit or in a multiple Type 400 mounting. Power Source 1 (Phantom Power) provides power on the transmit and receive wires of the S/T interface. Power Source 2 (dedicated pair) provides power over an additional pair of wires. In both cases, the NT-1 monitors the S/T interface for an overcurrent fault condition. In the event of such condition, the NT-1 automatically limits the amount of current being delivered until the fault condition is removed.

The S/T interface of the NT-1 may have as many as eight passive bus devices connected to it. These connections may be in one of the following two modes: multiple terminals within 50 meters of each other over the full 15-meter range of the interface, or multiple terminals of arbitrary spacing within 200 meters.

Termination impedance options include the U.S. standard of 100 ohms or the European unterminated configuration.

FEATURES

- Compatible with 2BlQ basic rate services and standard ISDN terminals
- Provides 2B+D basic rate service
- Conforms to ANSI Tl.601-1991 and ANSI Tl.605-1989
- Supports multiple switch vendors: AT&T 5ESS, NT DMS-100
- Mounts in Type 400 multimount shelf
- Available in stand-alone unit
- Supports S/T bus powering sources
    - Phantom power
    - Dedicated pair
- Provides automatic overcurrent protection for S/T power
- Selectable terminating impedance; supports U.S. or European standard configurations ['100 (omega) or unterminated]
- Allows long and short passive bus arrangements
- Provides U interface lightning protection

B4.B. AT&T

National Distributor: Anixter

Tel. 800-264-9837

AT&T has developed a new more compact and lower cost NT-1 designated L-230. No details were available in time for publication.

B4.C. MOTOROLA NT-1D

Motorola
500 Bradford Dr.
Huntsville, AL 35805-1993
Tel. 205-430-3000 Fax 205-830-5657

The Motorola NT-1D provides the interface between the telephone company ISDN network and the user's terminal equipment. The NT-1D is installed between the central office U interface and the customer premise S or T interface.

The NT-1D is a fully compliant 2BlQ Basic Rate network termination 1 (NT-1) unit as described in ANSI specification Tl.601-1991. The NT-1D converts a 2-wire echo-canceled 2BIQ U interface line code to a 4-wire alternate space inversion code. It supports both point-to-multipoint configurations.

FEATURES

U Interface

- U interface compliant with ANSI standard Tl.601-1992
- Performs embedded operation channel (EOC) 2B+D loopback
- 18,000 foot operating distance on 26 AWG
- U interface metallic termination (sealing current)
- Remote activated quiet mode and insertion loss tests in maintenance model
- Local power loss “dying gasp” notification
- Surge protection
- Warm-start activation

S/T Interface

- S/T interface compliant with ANSI standard Tl.605-1991
- Supports multiple wiring configurations
- Receiver and transmitter can be terminated by the integral 100 ohm resistors

General

- Desktop or wall mount
- Automatic remote (backup) power pickup
- All options and connects readily accessible on rear of unit
- All interfaces use standard RJ-45 jacks
- Wall transformer and U interface cable included
- Line ordering information not required

B4.D. NORTHERN TELECOM

Telnet to Digital Velocity BBS

919-992-0407 for ISDN and 9l9-992-3059 for Analog

The ANSI standard 2BlQ network termination 1 unit incorporates an ANSI standard interface for the subscriber loop and one for the customer interface bus, known at the S/T bus. The former allows the subscriber of U-loop to connect the NT-l to the ISDN line card in the telephone company central office, while the latter interface connects the NT-1 unit to the customer's terminal equipment at the workplace.

The NT-1 unit supports the ANSI standard 2BlQ U-loop interface. Utilizing the NT-1 user's ISDN terminals can access ISDN Basic Rate service offered by compatible telephone company switches. Basic Rate service provides clear transmission of circuit-switched voice or data communications or of packet-switched data communications.

The stand-alone NT-1, which terminates one ISDN subscriber loop, is designed to handle as many as eight customer voice and data terminals. A maximum of two voice terminals is allowed, as well as up to six D-channel packet-switched data terminals. Users may customize the system by dedicating one NT-1 to one person, providing voice, circuit-switched data, and packet-switched data services. Or, two users could each have voice and packet-switched data services. Or, two users could each have voice and packet-switched data, allowing one NT-1 to be shared by both users. Indicator lamps provide the user with U-loop and S/T bus synchronization. The NT-1 provides a diagnostic loopback point from the central office if trouble should be encountered in the U-loop. For additional convenience, the standalone NT-1 can be wall-mounted to maximize desk space.

For installations where desk space is at a premium or backup power is required, modular NT-1s allow multiple NT-1s to be conveniently mounted in racks in wiring closets. Up to 12 NT-1 cards can be mounted in an NT-1 module, which provides status with LED indicators as to the operability of each card. Diagnostics from the NT-1 module to the attached terminal can also be initiated and monitored.

Its modular design gives the rack-mount unit extra flexibility when it comes to configuration. Easily mounted on a wall or equipment rack, the modular NT-l may be powered either by optional Northern Telecom power modules or by a user's own power source. Each unit incorporates LED indicators to provide information on power status, as well as the status of both the subscriber loop to the network and the line from the NT-1 to the user terminal. Optional battery modules provide backup power to the NT-1s and ISDN terminals in the event of a power failure.

FEATURES

- Links central office provided U-loop with customer premise S/T bus for ISDN terminals
- Provides ANSI standard interfaces for the subscriber U-Ioop and the S/T bus
- Compliant with National ISDN-1 standards for 2BlQ signalling from central office to NT-1
- Stand-alone NT-1 supports a signal ISDN loop, with up to eight terminals per S/T bus
- Rack-mount NT-1 useful for localized NT-1s in wiring closet, providing battery backup, improving diagnostics, and eliminating wiring at users' desktops
- LED indicators give status of the subscriber loop, S/T bus, and power
- Stand-alone unit can be wall-mounted; modular unit(s) can be wall-mounted or rack-mounted
- Provides transmission performance monitoring and subscriber loop maintenance by the network.

B4.E. TONE COMMANDER

Tone Commander
11609 49th Place West
Mukilteo, WA 98275

Tel. 800-524-0024, 206-349-1000 Fax 206-349-1010
http://www.halcyon.com/tcs/

Tone Commander ISDN network termination (NT-1) units convert the ISDN network U interface to an S/T interface to connect to local customer terminal equipment. Tone Commander NT-1s are compatible with any central office that supports the ANSI standard 2BIQ U interface; they are also compatible with both national and custom ISDN standards. Desktop, wall, and rack-mount versions are available, as well as several different powering options.

The Right Power for Your Application

The NT-lU-220TC supports terminal equipment (typically ISDN telephones) powered through the NT-1. In a stand-alone configuration, power is provided by an in-line desktop power supply; in a rack configuration, power is provided by the rack, with battery backup options available. Self-resetting short circuit power protection is provided.

For installations that have locally powered terminal equipment (typically terminal adapters), the NT-lU-100TC stand-alone unit and NT-lU-110TC rack-mount card provide lower-cost alternatives. Low power consumption allows easy backup by any commercially available uninterruptible power supply (UPS).

Convenient Rack Mount Options

The NT-1-220 rack holds 12 NT-1U-220 units and can power up to 24 terminals in addition to the NT-1s. The optional NT-1 200 battery backup allows up to four hours of operation during a power failure; battery capacity may be doubled with an additional NT-1-200 add-on battery.

The NT-1-110 rack holds 16 NT-1U-110TC cards and includes an in-line power supply to power the NT-1s. This arrangement allows both the lowest cost and smallest space installation for high-density data applications.

Easy Setup, Reliable Operation

Termination selection is easily set with a single externally accessible slide switch. No timing selection is required in most installations. Status indicator lights allow quick verification of correct installation and operation. Self-resetting surge protection provides secondary lightning protection on the telco U interface to ensure reliable, uninterrupted service.

Tone Commander NT-1s have been used with virtually all ISDN terminal equipment and applications, in every central office environment.

Features

Tone Commander NT-1U-100TC

- Two status indicator lights
- Switch selectable S/T termination
- Supports National ISDN electrical and loopback testing modes and provides ANSI Tl.601-1992 DC signalling test modes
- Power supply included
- Shirt pocket-size desktop unit

Tone Commander NT-IU-220TC

- 4 status indicator lights
- Switch selectable S/T termination
- Supports National ISDN electrical and loopback testing modes and provides ANSI Tl.601-1992 DC signalling test modes
- Manual reset switch
- Desk top, rack, or wall mountable
- Packages dimensions and connector arrangements allow installation in Tone Commander NT-1-220 Rack or AT&T NTM-200 Rack
- Supports ANSI subchannels SC1, SC2, and Q multiframe communication

- Optional stand-alone power supply provides 16 watts power to terminal equipment; ANSI standard PS1 and PS2 configurations
- Self-resetting short circuit power protection

Tone Commander NT-1 110 Rack

- Rack holds 16 NT-1U-110TC cards
- Power supply included
- 19" or 23" rack or wall mounting

Tone Commander NT-1 220 Rack

- Rack holds 12 NT-1U-220TC units
- Integral power supply provides 120 watts for terminal equipment
- 19" or 23" rack or wall mounting
- Optional battery backup available

B4.F. IBM NETWORK TERMINATOR EXTENDED

See Address Above

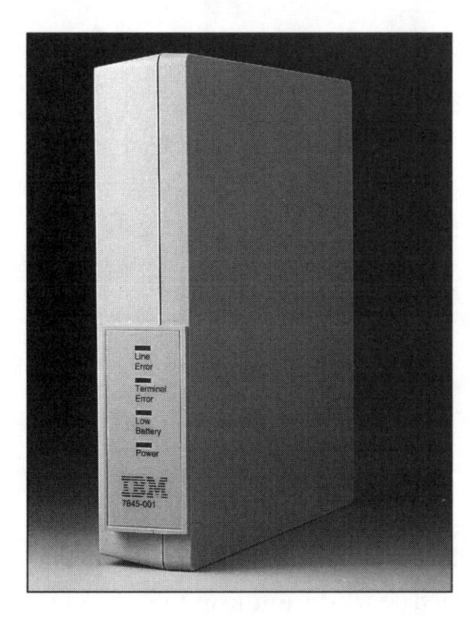

Should you need to set up multiple devices on a single BRI line IBM does sell a product, called Network Terminator Extended, that will give you an option for using both B-channels on a single BRI line.

Dividing up the bandwidth this way could be beneficial to users who would like to have one B-channel available for an analog telephone, facsimile, or other similar type of equipment.

With the NT Extended, you'll have the security of knowing you can use your telephone equipment even if an outage disrupts the electrical power to your home or office—exactly as if it were conventional analog service provided and powered by the telephone company central office. The NT Extended comes with a rechargeable battery capable of providing backup to your telephone sets for up to 18 hours of stand-by capability.

Features

In extended mode, you'll be able to use many custom calling capabilities similar to those offered as options with regular analog service by the telephone companies, including:

- Speed dialing
- Redial of last number dialed
- Repetitive redialing of the last busy number
- Return of the last incoming call
- Call hold
- Call retrieve
- Three- or six-way conferencing
- Call waiting

Highlights

The IBM ISDN Network Terminator Extended enables remote PCs to approach the efficiency of a locally attached workstation using ISDN while supporting full analog telephone services by providing a wide range of features and benefits including:

- Fully programmable ISDN network terminator functions
- Digital communications at 64 Kbps over one of the two B-channels furnished by ISDN BRI service, into an IBM Personal System/2® (PS/2®) or compatible PC equipped with an IBM WaveRunner digital modem
- Extended function that allows the other ISDN B-channel to carry analog voice communications using existing telephone equipment
- Full range of custom calling features, like those offered by telephone companies, easily enabled by programming using a standard touch-tone telephone
- NT-1 capability to pass both B-channels and D-channel to the remote workstation
- Rechargeable battery that provides backup power to the analog telephones during electrical power outages
- Compatibility with AT&T 5ESS, Northern Telecom DMS-100, and National ISDN-1 (NI-l) switches

These features give you a cost-effective solution access from the office or home office to corporate and suppliers' databases and transaction systems, plus access for telecommuters to corporate information systems facilities at efficient speeds.

Real/Time Communications

http://www.realtime.net/

512-451-0046

6721 N. Lamar, Suite 103
Austin, TX 78752

sales@realtime.net

Real/Time supports dedicated ISDN connections at either 64 Kbps or 128 Kbps.

Zilker Internet Park

http://www.zilker.net/zilker/zilker.html/

1106 Clayton Ln., Suite 500
Austin, TX 78723

Offers on-demand dedicated ISDN accounts.

Washington

Cortland Electronics

http://www.cortland.com/

800-877-0792 x61 (enter extension at tone)

206-217-0158

mike@cortland.com

Dial-up and dedicated ISDN connections at 64 Kbps and 128 Kbps.

Pacific Rim Network, Inc.

http://pacificrim.net/

800-591-2757

PO Box 5006
Bellingham, WA 98227

Offers 56 Kbps single PC and LAN dial-up.

Wolfe Internet Access

http://www.wolfe.net/wolfe/wolfe_info.html/

800-965-3363

206-812-4000 and Select “2”

support@wolfe.net

Offers dedicated 64 Kbps Internet access.

This appendix is intended to provide you with a short list of potential sources for additional peripherals, software and hardware, to work with your ISDN equipment. See chapters 4 and 5 to determine which of the following types of products will be required for your implementation.

COMMERCIAL TCP-IP PACKAGES

Company: Microsoft

Products: Windows for Workgroups with TCP/IP add-on, Windows 95, Windows NT

URL: http://www.microsoft.com

Company: Frontier Technologies

Product: SuperTCP Suite

URL: http://www.frontiertech.com

Tel: 800-929-3054

Email: info@FrontierTech.com

Company: FTP Software, Inc.

Product: OnNet for Windows Version 2.0, InterDrive95

URL: http://www.ftp.com

Tel: 508-685-4000

Email: info@ftp.com

Company: InterCon

Product: TCP/Connect II for Macintosh Version 2.2

URL: http://www.intercon.com

Tel: 800-468-7266

Email: sales@intercon.com

Company: NetManage

Product: Internet Chameleon

URL: http://www.netmanage.ocm

Tel: 408-973-7171

Email: sales@netmanage.com

Company: SPRY

Product: Internet in a Box

URL: http://www.spry.com

Tel: 800-777-9638, x201

Email: info201@spry.com

Company: TGV

Product: MultiNet for Windows

URL: http://www.tgv.com

Tel: 800-TGV-3440

Email: sales@tgv.com

SHAREWARE TCP/IP PACKAGES

TRUMPET WINSOCK

FTP Site: ftp.trumpet.com.au

Location: /ftp/pub/winsock

File: twsk21f.zip

FTP Site: ftp.acsu.buffalo.edu

Location: /pub/misc/ppp/winsock/trumpet

File: twsk21f.zip

FTP Site: utkux1.utk.edu

Location: /pub/dialup/packages

File: twsk21f.zip

MACPPP

FTP Site: merit.edu

Location: /pub/ppp/mac

File: macppp2.0.1.hqx

FTP Site: hub.ucsb.edu

Location: /pub/networking/mac

File: macppp2.0.1.hqx

FTP Site: nis.nsf.net

Location: /internet.tools/ppp/mac

File: macppp2.0.1.hqx

WINDOWS 3.X COMMUNICATIONS PORT SOFTWARE DRIVERS

KingCom
E Ware
145 West 28th St., 12th Floor
New York, NY 10001-6114
phone: (800)892-9950

TurboCom/2
Pacific CommWare
180 Beacon Hill Ln.
Ashland, OR 97520
phone: (800)856-3818, (503)482-2744

HIGH SPEED SERIAL CARDS

IBM COMPATIBLES

I/O Professional
Siig, Inc.
6078 Stewart Avenue
Fremont, CA 94538
phone: (510)657-8688

MACINTOSH

Hustler
2 port High Speed Serial Card
Creative Solutions, Inc.
4701 Randolph Road
Suite 12
Rockville, MD 20852
Phone: (800)367-8465
Phone: (301)984-0262
Fax: (301)770-1675

VersaLink
4 port High Speed Serial Card
Advanced Logic Research
Phone: (214)243-8700
Fax: (214)243-4280

EVOLUTION OF ISDN

TRANSMISSION DEVELOPMENT FROM SIGNALS TO ANALOG

This appendix covers the development of transmission systems and switching systems. We will discuss the change up through analog methods and then the conversion to a digital basis and, finally, the beginnings of ISDN.

The first transmission systems involved our physical senses directly. Sight, hearing, and smell operate at greater distances and, thus, are more directly relevant to the discussion of transmission systems. Hearing is the reception of data in the form of sound. The creation of the data involves something creating sound waves. Thus, a shout may be able to be heard at a distance. We are able to hear (and interpret) sounds as language. Natural items (such as birds taking flight) or human-made noises (such as automobile traffic) act as warnings. Sight uses light as a transmission medium. The eyes are receptors, and the origination is created by reflected, or produced, light from objects. Smell is no longer used much by people, but, at one time, it identified friend and foe.

All of these natural senses have limitations for distance. It is natural, therefore, that the first human-augmented transmission systems involved methods of extending the distance that could be covered. In what ways was this possible? This was done in two primary ways: the use of relays and the use of storage. A relay system extends distance in an additive manner. A person can shout to another 500 feet (as an arbitrary limit) away. That person can shout to another 500 feet away. Thus, a relay of three people can cover a distance of 1,000 feet. The same can work with sight—but the distances are longer. A signal fire (at night) may be seen a mile away (further depending on topography). From that point, another signal fire may be created to extend the distance.

Storage allows for delayed transmission. A message can be written down and physically transported to another location where it is read. This extends the distance possible and is also a form of indirect translation (from thoughts to written symbols). A courier with a good memory may be told a message and be prepared to repeat it after travelling to the desired destination.

Relaying and storage allow extension of distance limitations. However, storage methods rely on physical transportation, which may be slow. Relays are limited in the number of discrete signals and will require large numbers of relay sites. Assume that a relay limitation is 1 mile. To be able to communicate to each point in a 5-mile by 5-mile area, 25 relay points are needed. Nevertheless, relays and storage are still the primary methods used for transmission of data over long distances. Original methods such as the "Pony Express" courier system and relays of smoke signals are slow in comparison to modern methods, but the concepts remain much the same.

The search for added speed first worked toward more efficient use of human labor and ingenuity to lengthen distance and decrease transmission time. Harnessing electricity allowed direct routing of information near the speed of light. In one respect, this is not different from lighting signal fires from the tops of mountains. In other respects, however, it is much more reliable and able to decrease the amount of human intervention in the process—a bottleneck for speed and content.

The transmission of electrical current was, at first, only a binary signal—used to provide power to the other end of the line. The power existed or not. The information contained was thus one of whether the circuit was unbroken or not—a useful item of information but still a binary signal. The next step was to encode information on the line. Morse code (see Figure G.1) was developed to send coded information on an electrical circuit. The pattern of interruptions in the current could be deciphered into a much more complex signal. At this point, other transmission methods began to be developed to allow transfer of data without a predefined circuit (such as that provided by an electrical circuit). The use of radio waves provided this capability.

In a manner of speaking, transmission networks have returned to the beginning, since digital transmission is a refinement of those early Morse code signals. The desire existed, however, to be able to convey more complex signals; equipment that could do this in digital form did not exist. Thus, analog information was sent. Analog data are a continuously

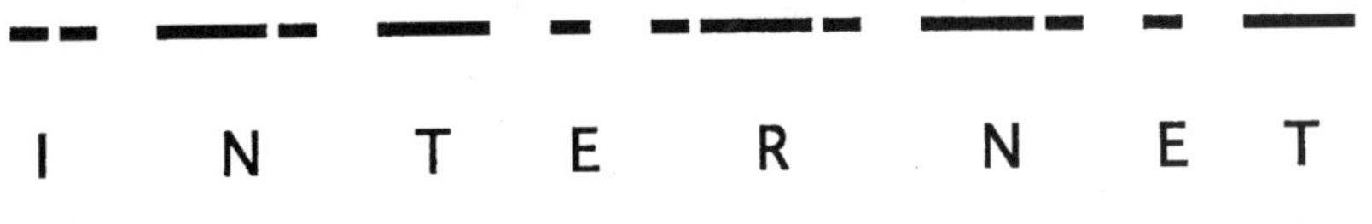

I N T E R N E T

Figure G.1 Example of Morse Code.

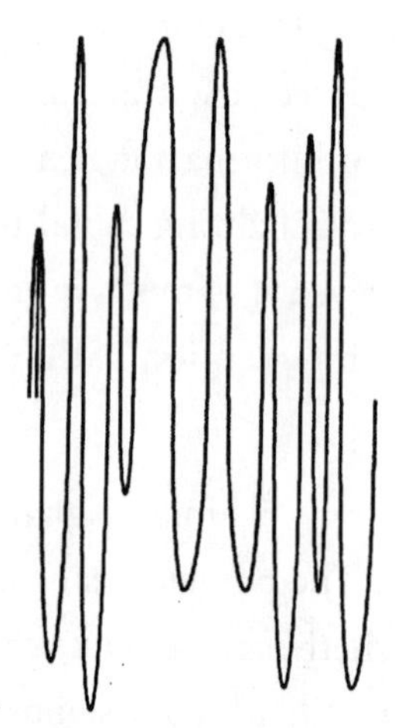

Figure G.2 Description of a sound wave.

changing stream of information. This fluctuation in frequency and amplitude is translated from one form (sound) to another form (electrical currents or broadcast signals), which can be transmitted, received, and translated back into a form that can be used by the recipient. Figure G.2 shows a description of a sound wave. This pattern describes equally the sound recording or the electrical impulses. Thus, use of analog transmission methods for sound is a simple and direct extension of manipulation of electrical currents.

Telephony, or "far sound," was developed to carry complex information across long distances. It uses direct transmission paths to allow one endpoint to communicate with another. However, it is desirable to be able to change the path to let different people talk with each other. This is the topic of switching.

THE DEVELOPMENT OF ANALOG SWITCHING NETWORKS

Switching would not be necessary if a direct path existed between all endpoints that need to communicate. This is not feasible except in special, isolated, circumstances. The "red phone" at a security location is one example. A direct path exists between one instrument and another. If 5 people wanted to talk with each other, however, it would require 10 direct connections to allow the full interaction. Ten people would need 45 direct connections. This continues to increase in a nonlinear fashion with additional connections. Thus, a full interconnecting set of direct connections is not feasible.

The first switches used human operators to change the connections from one party to another. Each endpoint has a direct line to the switching station. The operator then can manually connect one endpoint to another by supplying a junction connection between the two endpoints. Ten people can speak with one another by use of 10 direct connections to a common switching center. This was called a switchboard switching center and was used in the United States until the early 1980s in some areas of the country. Long distance service adds the need for the switching centers to be connected together. Forty people, served equally by two switching centers, could therefore be fully interconnected by the use of 60 direct connections. This, as seen in Figure G.3, is done by having 20 direct connections to each switching center and an additional 20 connections between the switching centers.

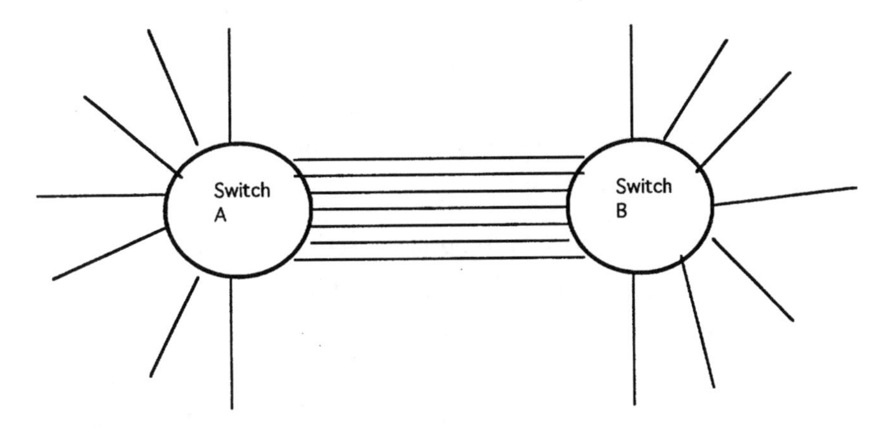

Figure G.3 Full capacity switching.

In reality, it is rare that the switching centers would be connected by the same number of connections as potential users. Instead, networks are engineered such that the number of connections is sufficient to allow for "normal peak" levels of service. When use exceeds these amounts, some percentage of users will receive feedback that basically says that the network is busy and to try again later. This is different from a "user busy" situation where the other party is already using the circuit connection for another call.

ANALOG SIGNALLING METHODS

Signalling information, or the specification of with whom one wants to speak, is straightforward in a switchboard operation. A simple request, or signal, is given to the operator. The operator enters into conversation with the person placing the call, who then tells the operator the desired endpoint. The operator makes the necessary connections.

Elimination of the human intervention lowers costs and increases speed of connections. In order to do this, automating the actual switching of the circuits is necessary, as well as devising a system of signalling that the equipment can use to determine the appropriate switch connections.

Signalling is part of the process to tell the network what ends must be connected. The first part of this process is to indicate need of service. For an analog phone, this is done by removing the receiver from the cradle (or by pressing a button indicating need of service). This causes a simple completion of the electrical circuit from the user to the service location. A monitoring process will be responsible for checking the status of each circuit that is part of its monitoring group.

Once the monitor has noticed a request for service, a "dial tone" is applied to the line and a signal processor assigned to the circuit. This dial tone is an audible (potentially translated to other signal means) indication that the line is in service and available for use. A timer will usually be associated with this such that an available circuit will not take up network resources if it is not actually being used (this is usually a loud warning alert sound with a possible added voice message).

The circuit is now available and a signal processor is waiting to process any information given by the user. The signal processor is usually dedicated because timing is important for the *pulse* dialing originally used on analog lines. Pulse dialing incorporates breaks in the circuit in a manner very similar to that of use for telegraph use. Each break in the circuit is of a specific length (there are usually ten pulses per second). Additionally, the total number of such breaks, without added delay, makes up the actual signal element. Thus, one break in the line is an indication of a signal "one" to the network. Ten breaks (within a total amount of time not exceeding a second) indicate a signal of "zero" to the network. This method allows the use of a combination of ten unique digits to be relayed to the network.

The other method of signalling used by individual circuits is that of *tone* signalling. Voice information (or speech) over an analog circuit will take place over a particular bandwidth. This bandwidth is a reflection of the range of frequencies needed to adequately convey speech information. It is possible to put additional information outside of this bandwidth onto the circuit. This is called "out-of-band" signalling. We will see that this term is also used in digital signalling methods, but for a slightly different method. In both cases, out-of-band means that there is a different path for the signalling information than for the actual data being transferred.

In most North American systems, the tone signalling is placed "in-band." This is why the user can hear the tones that are selected. Each tone is actually made up of a combination of two frequencies. This unique combination of frequencies for each tone makes it unlikely that the same tones will be produced by the human voice. Tone signalling has the main advantage that it can be produced, and accurately detected, in less time than pulse signalling. A single pulse signal takes a second for transmission and decoding. In this same period, it is possible to have ten tone signals. This means that the signal processing resource may be freed more quickly.

ANALOG SWITCHING

Once the local service provider has been notified of a request for service and given sufficient information to set up a connection to the desired endpoint, we have entered the area of switching. Switching is the process of providing a connection between two or more points over a circuit that is transient in nature. In other words, the circuit is in place only during the existence of the call.

Early analog switching relied on electro-mechanical switches. One such was called a step-by-step switch. Step-by-step switches (introduced for long distance use in 1926) are still in use in some rural areas of the country. They have the advantage of being very simple; maintenance is low, and the need for servicing is very infrequent. Each signal element causes one set of connections to be completed. When the final signal element has been received, the connection is complete (or, at least, that stage of the connection). Although very economical, cross-connect switches do not allow any advanced features that are now considered a necessary part of network service.

In the case of remote service, the endpoints are each served by a separate network node. That is, the electrical circuit for each person to be connected is physically present between the user and a different service node. This means that there must be some way for the nodes to communicate the signalling information between the service nodes.

DIALING PLANS

A dialing plan allows the signalling information to be routed appropriately. Currently, in North America, this is broken into a prefix code, area code, switch identification, and user identification. The same types of information are used within the dialing plan for each country, but the exact form will vary.

In North America, a prefix code of 1 will indicate long distance service. A prefix code of 0 indicates operator intervention (although current needs have expanded the definition of this to sometimes not have a human operator). Each prefix defines the meaning of the digits that follow. The code 911 can thus be used as a special code for emergency services in all areas. A code of 011 can be used to indicate international directly dialed codes. (In this case, the next numbers will indicate a country code, optionally followed by a city code, then local number.)

For most users, the 1 is the prefix most often used. The next set of three digits constitute the area code. The area code is used by the national switching network to provide service between different regions of the country. The next three digits identify a particular switch that gives service to up to 10,000 users (10,000 endpoints can be identified by four digits). Thus, there are four stages to setting up a long distance call. These are access to the long distance network, establishing a connection to the region desired, identification of the switching system that provides service to the user desired, and final connection to the appropriate endpoint.

Note that each dialing plan can be modified as needed to provide for growth in the system size as well as new features. The 800 "area code" provides access to a service rather than a geographical region. The 011 prefix was not an original need of the network but, rather, replaced what used to be an operator-provided service. So far, the dialing plan, or numbering plan, has been sufficient to provide *backward-compatible* service. In other words, the changes to the dialing plan allow old services to be provided without change in procedure (although a new area code may be allocated—changing the full, long distance, identification number of a subscriber). Only new services require changes in access methods. It is difficult to know whether this can continue, considering the huge expansion of uses of the telecommunication networks.

Long Distance Signalling

We talked about local signalling to provide information to the local service node as to the endpoint desired. It would be possible to repeat this information, in the same form, from the local service node to other nodes.

However, this information is no longer sufficient as the connection is established from one location to another. In one respect, it is more information than needed. In another, it lacks information.

For example, once the connection has been made to the region addressed, the area code is no longer needed. However, for reverse charging purposes, the origination information (known to the original service node) may be needed.

We have used the term "node" several places in our discussion of network needs. A node is a place where connections come in and go out. It is a place where additional switching will occur. In the case of completely local service (the originator and called party are both serviced by the same switch), only one node is involved and only one switched connection needs to be established. In the case of other local service (not officially long distance), a second connection must be made between the two switches being used for the connection. Thus, two nodes are in use. There is a switched connection between the originating user and the switch being used by the called party. There is another connection between the remote switch and the called party.

When long distance service is required, the number of nodes involved will be unknown. As an example, say that a call is to be placed between someone in Wichita, Kansas, and Madison, Wisconsin. If there is a lot of traffic between the two places, it is possible that there will be a single long distance circuit that provides direct connection between the cities. This will mean that four nodes are in use: the local and remote switches that provide the direct circuits to the endpoints plus two long distance nodes that connect the cities.

It is more likely, however, that regional switches will be used. The traffic will be *routed* based on traffic needs. There may be a circuit available between Wichita and Kansas City. Kansas City is then connected to Chicago and Chicago connected to Madison. However, if a lot of telephone traffic already exists between Wichita and Kansas City, the call may be routed through Omaha—or even Denver.

It is also likely for each area code to have a specific node that traffic will be routed through first. Thus, Wichita's area code (316) has a specific node through which all long distance traffic (outside of the area code) is routed. This node then makes connections through regional network nodes until it reaches Madison's area code (608) and then is routed to the final service node for the called party.

All of this routing requires signalling information to be carried through the network. This may be done in-band or out-of-band. Before the connection is completed, the circuit may be used exclusively by the network (the users cannot actively speak over the circuit, so why not the network). A single frequency method (usually at 2600 Hertz) can use the same bandwidth as the voice circuit. The signalling information may also be translated into an out-of-band signal that is carried with the same circuit. Technically, this is out-of-band but it has characteristics that associate it with in-band since the same circuit carries the information. A 4 kHz bandwidth is allocated for voice circuits. Voice usually only uses the bandwidth between 500 Hertz and 3500 Hertz. Thus, the frequency between 3500 Hertz and 4000 Hertz is available for the network.

It is also possible for the network to use completely out-of-band signalling. One such method is called Common Channel Interoffice Signalling (CCIS). This allocates a separate circuit between the two switch nodes for signalling uses. This is more efficient than in-band signalling and prevents any type of interference between in-band use and signalling needs (particularly for additional signalling needs while the call is in use—such as call waiting). It also provides an additional degree of switch security because this special circuit may be accessed only by the network.

Use of Long Distance Trunks

We have talked about long distance connections as if they were single circuits available for remote connection. This was certainly true in the earliest days of the telephone network. However, it is very inefficient to allocate a single line to a single call. Instead, a trunk line is allocated. This trunk line allows a large number of circuits to be allocated to the same physical circuit.

We said that each analog voice circuit is allocated a 4 kHz bandwidth for use for voice. If a trunk can carry a frequency bandwidth of 4 MHz, then 1,000 separate calls can be carried over the same physical circuit. This process is called frequency division multiplexing (FDM). Actually, due to overhead and the need for signalling channels, a 4 MHz line would not be able to provide for 1000 separate calls—but the basic description still is true.

DIGITAL TRANSMISSION USE

In an electrical communication network, there is no real difference in form between analog and digital use. Both are composed of an electrical current that varies in voltage. The difference (as seen in Figure G.4) is how the signal is interpreted. An analog signal, which varies continuously, has a large number of values that can be interpreted from the signal. A digital signal has a fixed set of values that are used within the signal.

Stereophiles will argue over the merits of analog recordings versus those of a digital medium. They will usually claim that analog is much better. They are correct that analog provides much more information and, since the signal is a direct conversion of audio to a recorded medium, will produce a

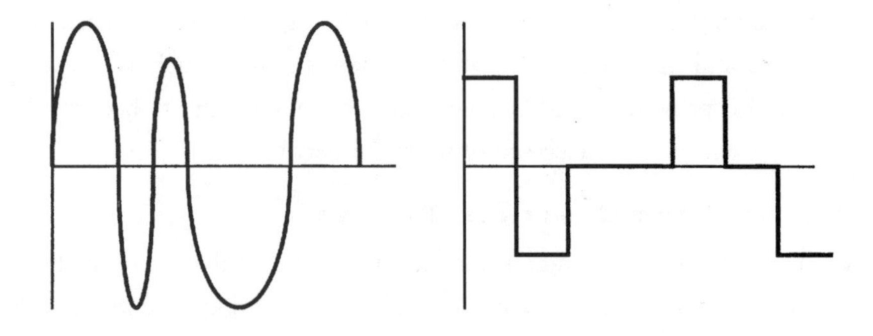

Figure G.4 Analog versus digital forms.

more accurate representation of the original signal than digital is ever likely to provide. In theory, digital information can provide the same amount of data that analog can—at least, to the point where the human ear can discern the information. In practice, however, digital encodings do not (primarily for space and economical reasons) bother to store that much information.

However, analog transmissions and recordings both are susceptible to distortion. An analog recording, recently made and reproduced with very good equipment, will give an excellent reproduction of the original sound including any overtones and other distinctions. A worn physical recording, however, will introduce many variations from the original recording.

In a similar manner, analog transmissions are susceptible to distortion also. These arise from equipment problems as well as inherent loss of information over distance. Because analog information is sampled continuously, every such change in the data will degrade the signal. On a voice signal, the human ear will compensate to a large degree and will replace the lost information according to context and memory. If the circuit is particularly bad, the static may be such that information is completely lost. For the most part, however, analog transmission is adequate for voice transmission.

When analog transmission is used for data (non-voice applications), this distortion of the signal will limit the amount of data that can be carried accurately. Thus, the signal is limited to a discrete number of signal levels rather than a continuous analog data stream. Error correction methods can boost the possible speeds but distortion will inevitably cause limits.

Digital transmission methods expect distortion. As seen in Figure G.4, a digital signal is composed of levels. Each level is actually a range of values, distortion is much less relevant. Say that a voltage level of one positive volt is meant to mean a binary value of 1 and a voltage level of one negative volt indicates a binary value of 0. The voltage can actually vary by a volt (or even more if the medium allows greater voltage values) and still be interpreted correctly. Digital sampling—checking the value at discrete time intervals—directly allows for the possibility of distortion of the signal.

Thus, the signal may remain intact even if distortion occurs. Also, since the *exact* signal is known, the transmission line can use digital repeaters that can eliminate the distortion. Say that we send a signal from point A to point D, with switching nodes at B and C. When the signal reaches point B, some distortion has taken place but not enough to hide the original data. The repeater can recreate the signal without the distortion before forwarding the data to node C. Data are lost only when transmission problems become so bad that the data values are lost between nodes.

Another advantage in digital transmission is that signals can become more specific, or complex, by having longer signal sequences. This is possible in analog transmission but, due to distortion, the speed is more greatly restricted. For example, the opening bars of Beethoven's Ninth Symphony could be used to indicate the use of call forwarding. The same thing could be achieved in digital transmission by the use of a sequence of digital information that might take place in a few milliseconds.

DIGITAL SWITCHING

Digital switching provides for the same basic operations as does analog switching—it routes data (including voice) from one point to another. However, because the form of the data is different, there are different operations available to digital switching. The first variance is that two types of switching are available. The next involves methods of multiplexing. Another is involved with handling of the actual data.

CIRCUIT-SWITCHED VERSUS PACKET-SWITCHED

The two switch variants are called circuit-switched and packet-switched. Circuit-switched, the only form available for analog information, forms a continuous link from one point to another. Once the link is established, it is not normally rerouted. If the link is taken down (voluntarily or not), full call signalling and processing are needed to reestablish a connection.

Packet-switched networks may be dynamic because the packet contains addressing information within the segment of data. If a particular connection becomes overloaded with traffic, another route may be established without user intervention. A disruption in a connection may also be recovered. It is also possible to have the connection tariffed according to use. For a circuit-switched connection, the link is dedicated and, thus, may be charged to the user on a continuous basis. In the case of a packet-switched situation, the network resources are shared. This allows for a variable rate that depends on the actual amount of data transferred. Thus, packet-switched connections are well suited for variable data transfers.

If the amount of data is relatively constant, a circuit-switched connection will be more cost-effective, partially because of tariffing but also because the address information within each packet forms an amount of overhead that may be (but is not necessarily) eliminated from a circuit-switched transfer.

MULTIPLEXING TECHNIQUES

We discussed frequency division multiplexing (FDM) earlier in conjunction with analog switching. This method provides shifting of frequencies so that a number of lower bandwidth channels may be carried on a high-capacity circuit. Digital information allows data to be carried based on time slot allocation. This method is called time division multiplexing (TDM). Figure G.5 shows the two methods. It is possible to combine the two methods—particularly with fiber optic transmission media. For example, it would be possible to send data over a fiber optic cable that varies according to frequency and then use TDM techniques to combine multiple channels on the same cable.

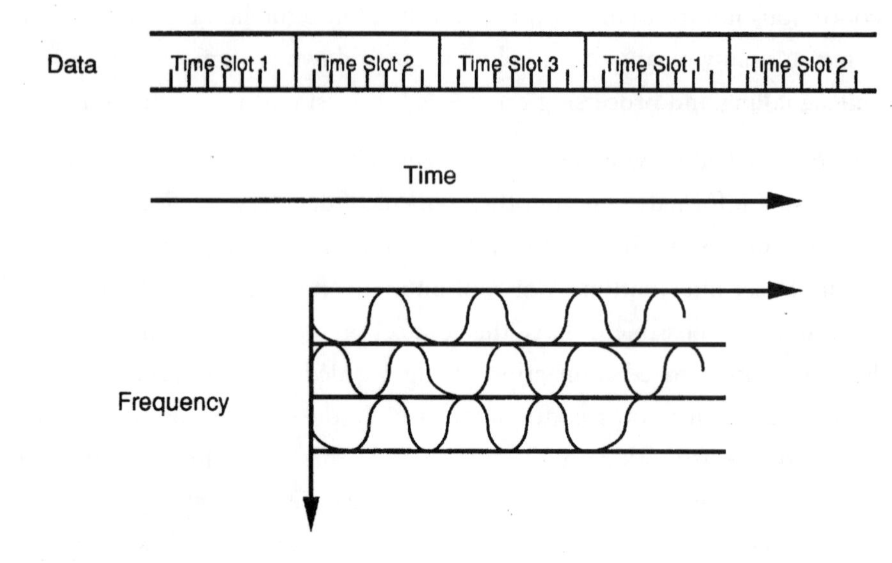

Figure G.5 Frequency and time division multiplexing.

It is normally much less expensive, more reliable, and faster to have each data stream on a fiber optic cable be made of a single frequency. This argues for use of binary, or digital, signals. Rather than a change in voltage, the presence or absence of the light can indicate a binary change of state. Plus, with the use of laser signal sources, the frequency can be kept within a very tight range. Thus, a cable can carry thousands of separate channels, each at a different frequency, and then each channel can use TDM methods to allocate the bandwidth.

DATA HANDLING TECHNIQUES

We discussed how segmenting digital data can allow use of packet-switching techniques. Even without packet-switching, digital data facilitates the use of network resources. Let us say that, within a switch, data must be switched from point A to point B (each of which may lead to another node or a user). Analog data, since they are continuous, must have a physical connection between the two points to keep the data intact. Digital data allows for the possibility of buffers. Figure G.6 shows a general outline of

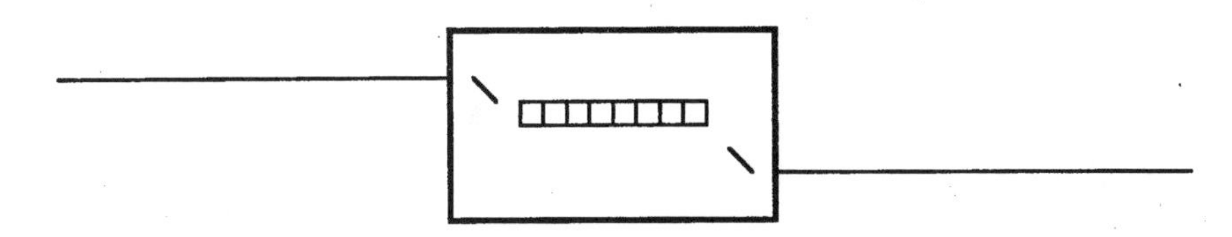

Figure G.6 Digital buffering.

digital buffering. As long as the buffer can be handled at a speed at least as great as the transmission speed, it is possible to eliminate the direct connection between the two switched points. This is similar in effect to the digital repeaters (or regenerators) mentioned earlier in this chapter.

DIGITAL SIGNALLING

Signalling is still required in digital switching systems. Like analog, it may be passed as in-band information or as part of a separate out-of-band signalling channel (channel is often used rather than circuit for digital use since the actual transmission link is less certain in form). In the United States, switched 56 Kbps service makes use of in-band signalling. It does this by using the high-order bit of every byte of data that is sent across the data channel. This means that, on a 64 Kbps channel, only seven-eighths of the bandwidth (or 56 Kbps), can really be used for data by equipment. ISDN, itself, is defined to use 64 Kbps channels so connections that must be routed through switched 56 Kbps service are a form of interworking. This situation is still prominent in the United States, although the other areas of the world do not have to contend with this potential problem.

A better solution (both from the point of data usage and for allowing full ISDN access) is to make use of out-of-band signalling. Signalling System 7 (SS7) is the ITU-T standard used for most inter-trunk long distance signalling for digital transmissions. Unfortunately, SS7 is not fully implemented in the United States, and it usually causes the need for interworking

ISDN with switched 56 Kbps service. SS7 can be looked at as a digital version of CCIS.

DIGITAL MIGRATION FROM THE TRUNKS TO THE HOME

Digital transmission and techniques were first used as part of the long distance networks to carry data from one network node to another. This was done for two reasons. First, it was possible for the long distance carriers' transmission methods to be changed without any change in equipment at the users' locations. Second, there was a great need for more efficient transmission mechanisms.

If you pick up your phone and place a call, you expect to be able to do so in the same way from day to day. When touch-tone dialing was introduced, people had to buy new equipment to make use of the new signalling method. They also had to change the way they used the phone. (Today, it is more common for people to not know how to use a rotary, pulse signalling, phone.) This was a change in equipment and usage. Such changes require time and are expensive for the user. Conversion to digital technology is a problem of the same nature.

However, changing the connections between long distance carriers did not require any change for the user. The long distance transmission carriers (in some countries, this is provided by a single PTT) could continuously make changes in their networks. Sometimes the changes were minor—such as changing echo suppressors (which are needed to prevent electrical signals from bouncing back on the circuit) or repeaters. Sometimes they were major. The change from electrical to microwave or fiber optic is really a major change of the network. In all such cases, the user was unaffected (except, we hope, for having a more reliable circuit). Digital transmission was introduced in the same transparent way.

Once digital transmission was introduced into the networks, it began to move toward the switching centers. Most of the ones that people use are located in their local phone company. Others, however, are located in businesses. In particular, private branch exchanges (PBXs) are used in larger corporate sites to provide internal call control and access, as well as to connect to the networks.

It was, and is, certainly possible for equipment to remain analog in nature and still make use of digital transmission. In fact, many manufacturers were reluctant to change technologies because the old methods were very reliable and already established. Still, any conversion adds to overhead and cost for the entire link. Thus, the next step in migration was to change the switches, including PBXs, to handle digital transmissions.

Note that, at this point, the user still didn't have access to (and didn't have to change to use) the digital data directly. It was possible to continue to use only proprietary protocols. In many instances, no special protocol was used over the digital lines. Instead, digital encoding schemes were used to translate the analog data directly into a circuit-switched digital stream.

However, the next step was to bring digital transmission all the way to the user. This presented many problems. The first problem was that of standardization. The next was to provide a method to migrate slowly from analog to digital equipment. This was needed to save the user costs in buying new equipment all at once. It was also needed because a network is a large investment of capital and material. As long as the changes were internal to the network, it could change one piece at a time without concerning users about how they were going to access the network.

STANDARDIZATION OF DIGITAL ACCESS

Why is standardization needed if digital access is extended to the user? There must be some common method to talk with the network. With analog signalling, signals pass only to the local network node. With digital signalling, most signals are handled by the local network node—but it is useful to pass some information all the way from one endpoint to another. If this is done, some standard must exist.

It is also useful to have a common standard to be able to have equipment that can work on any switch. In the United States, this is not fully the case currently although there are efforts to bring the various specifications into agreement. Still, there is enough commonality in the different specifications that equipment can, without an exorbitant amount of additional software, support the various specifications. This requires that users know just which variant is supported locally but still allows them to use common equipment.

It would have been possible to continue to use in-band signalling, such as is used with analog equipment. There are two drawbacks to this. The first is that it requires the network to constantly monitor the channel from the user's site. The second is that it will reduce the amount, and complicate the form, of the data transported over the channel. Thus, an out-of-band signalling method is preferable.

The final aspect of standardization to be considered is just who will be responsible for the standard. Every country has its own telecommunications standards body. (In some cases, their role is limited to choosing the standards to apply in their country, tariffing, and enforcing the use of the country's networks.) It would be completely possible for every country to have its own standard. However, if this were so, it would be difficult for one country to communicate with another. In the global community of today, this is not desired.

MIGRATION PATHS FROM ANALOG TO DIGITAL

The next important aspect of standardization is to provide a way to use existing equipment and networks at the same time as new equipment and networks are in use. This argues that it is necessary for basic service to use the same wiring that is already installed from the networks to homes and companies. This does not mean that other, new circuits may not be desired.

However, basic service must be provided without changes in the installed base. If the same circuits can be used for both analog and digital services, then one line can be converted without affecting the other lines. This is a form of migration.

Next, the standards must provide for the same services that are currently being used. Voice, or speech, is going to be of high priority. Analog equipment must still be able to be used within the new networks and standards. Conversion can be encouraged through tariffing structures, but the primary reason for conversion will be based on available equipment and applications that can make full use of the new technology. People change what they use, and how they use it, based on need or desire. The need can be forced on them but, without subsidies, this will be met with great resistance because it will spread the cost of the new technology directly to the user. It is much better to provide better service at lower costs. In this way, change becomes voluntary.

A

B

E

F

G

H

I

J

L

M

N

T